Gravity Does Not Exist

Vincent Icke

GRAVITY DOES NOT EXIST

A Puzzle for the 21st Century

Amsterdam University Press

Cover design and lay-out: Gijs Mathijs Ontwerpers, Amsterdam
Illustrations: Vincent Icke

ISBN   978 90 8964 446 6
e-ISBN 978 90 4851 704 6 (pdf)
e-ISBN 978 90 4851 705 3 (ePub)
NUR   924

Contents

Foreword

This is a Small Book about a Big Question, not a textbook of known physics. Or perhaps it's about a Big Opinion — or a small opinion, depending on one's perspective. It's a book about unknown physics. Every scientific fact was born as an opinion about the unknown, often called a 'hypothesis'. Opinion gradually becomes fact when evidence piles up. By perceptive and diligent work, it is

> *… possible to attain a degree of probability that quite often is hardly less than complete certainty. Namely, when the things that one has deduced from the supposed principles correspond perfectly to the phenomena that observations show us,*

as Huygens wrote. It has been so ever since, except that instead of 'supposed principles' we now say 'theory'. But what if there are *two* theories, each of which has produced a myriad of 'things that correspond perfectly to the phenomena' but that cannot be combined? One theory replaced the mystery of gravity by a precise picture of space and time. The other replaced the mystery of matter by a description of quantum particles that is so exact that some of its predictions have been verified to eleven decimal places. At the present time in our Universe, we may keep these two separate, each in its own domain: space and time for very large things, particles for the world of the very small. However, 13.8 billion years ago, these two incompatible theories referred to a single realm. Many scientists think that they can be united only by a minuscule group of hyper-specialists. I think differently. The mathematics of the ultimate answer will be as arcane as always, but that formulation will have to follow upon some original perception. Insight is freely distributed; all you've got to do is pick it up. I hope that somewhere a girl or boy will do so, because the generations of physicists who made the existing brilliant theories will soon be extinct. We will never understand the beginnings of our Universe until this puzzle has been cracked. That is why I hold the opinion that this is not just a big question, but the Biggest Question in physics of the 21st century.

The Process of Measurement

With measured tread, Johan Cornets de Groot climbed the many steps to the first tier of the tower. He did so solemnly, as seemed proper for his rank as burgomaster of Delft in southern Holland. De Groot had been invited to witness a physics experiment proposed by Stevin,[1] the Flemish engineer, polymath and private physics instructor to Maurits, Prince of Orange.

This is the way it happened in my imagination. What these two men actually did has not been recorded, except for the setup and outcome of their experiment. The year was 1585, in an era when the scientific acceptance of observations and experimental evidence was beginning to grow in the minds of the intelligentsia (the illiterate stonemason and the shipwright had always respected facts, of course). In our 21st century, surrounded at all times and in all places by the products of science, it is difficult to appreciate how radically new it was to conduct an experiment that brushed aside nineteen centuries of philosophical opinion and, indeed, to devise such a test in the first place.

High above the ground, the experimental apparatus was held ready: two leaden balls, one ten times heavier than the other, prepared by Simon Stevin of Brugghe. He was a scientist in the best modern sense of the word: his brain held a vast amount of knowledge; he was familiar with all the classical works on physics and mathematics known in his time; his own work advanced science and engineering; and he informed non-scientists about the wonders of the world — among them Maurits, Prince of Orange, for whom he composed a fat compendium of theoretical and practical physics and mathematics entitled *Wisconstige gedachtenissen* (Flemish for something between 'mathematical musings' and 'mathematical inventions').[2]

1   Simon Stevin (1548-1620), Flemish scientist.
2   Simon Stevin, *Wisconstige gedachtenissen*, Jan Bouwensz., Leiden 1608.

Stevin knew a lot, but he also understood that science is not so much about knowing as it is about searching. Of course it is necessary to be aware of the state of knowledge, but mostly to determine one's point of departure on a voyage into *terra incognita*, and possibly to get some idea about what direction to take in that immeasurable land. Like Galileo, Huygens, Newton[3] and others, Stevin was one of the founding fathers of science, known as 'natural philosophy' at the time. The trade was sometimes also called 'experimental philosophy', and that is the expression I prefer to use because of its nice sharp taste of active research.

In *De Weeghdaet* (literally meaning 'The Act of Weighing' but the implication is 'The Process of Measurement'), one of the chapters of *Wisconstige gedachtenissen*, Stevin reports on the experiment he conducted on that tower in Delft: dropping two leaden balls at the same time, one ten times heavier than the other, in order to see if — as Aristotle[4] had insisted nineteen centuries before — the more massive one would arrive first at the foot of the tower.

This experiment is almost always attributed to Galileo, but there is only anecdotal evidence[5] that he performed it, and then not before 1590. In any case, he did not publish his results; this is significant, because Galileo was always ready and eager to tell the world about his discoveries. In his writings, he merely describes a thought experiment, wondering what would happen if a light and a heavy stone were connected by a nearly weightless thread.

3 Galileo Galilei (1564-1642), Christiaan Huygens (1629-1695), Isaac Newton (1643-1727); Italian, Dutch, English scientists, respectively.

4 Aristoteles of Stagirus (384-322), Greek philosopher.

5 Michele Camerota, *Galileo Galilei e la cultura scientifica nell'età della controriforma*, Salerno Ed., Roma 2004, pp. 61-63.

→ The Oude Kerk in Delft in 2014, silhouetted against a Hubble Space Telescope image of the interstellar nebula NGC602. The tower is leaning a little. It is thought that Stevin conducted his lead-ball experiment here. The precise spot is unclear; it may even have been inside the church, where several suitable locations also exist.

In fact, Stevin's procedure was much more subtle than is reported in the tales about Galileo on the Leaning Tower of Pisa. To begin with, the release of the balls was witnessed by an independent observer who, given his social status, could not afford to endanger his reputation for being just and impartial. Furthermore, how would one determine whether or not the two balls arrive on the ground at the same time? No equipment existed to measure the time of the fall, and, after a drop of at least ten metres, the balls would be moving far too quickly to allow a determination by eye. Stevin placed a wooden board at the foot of the tower, put one of his aides next to it with his back turned towards the board, and merely asked him to tell whether he heard one thump or two. The servant reported

... that together they impact the board so equally, that their separate sounds appear to be a single blow.

Simply brilliant — which one of Stevin's contemporaries would have invented such a robust elegance?

It was a dramatic result, because Aristotle and his followers had always stated that objects fall more quickly if they are heavier. It's dramatic, because this type of *experimental* philosophy was based on the principle that fact takes precedence over opinion and authority, a principle that has enlightened the world ever since.[6] Before that time, the opinions of scholars soared high above the practical facts of mere craftspeople and engineers: if a philosophical dictum did not match a test, then so much the worse for the test.

Even today, Stevin's observation is a dramatic result, because it is a matter of life or death. In the summer of 2009, a man jumped into the Niagara River and let himself be carried over the falls, hoping to end

6 Later experimenters, even more subtle than Stevin, have performed a variety of increasingly precise tests in order to see whether the acceleration due to gravity depends on an object's mass, composition, or other properties: Eötvös, Dicke, Braginskij and others all found that the answer is no. Some of these experimental results are accurate to twelve or thirteen decimal places.

his life. He survived, however, wet but unscathed. Just as in Stevin's experiment, the water in the river dropped with the same acceleration as the man. When he arrived at the foot of the falls, he was surrounded by tons of water travelling with the same speed as he did, protecting him so well that he lived to tell the tale. Had he jumped down beyond the falls, he would have fallen 56 metres to hit the river at about 100 kilometres per hour. At that speed, water doesn't feel much softer than rock. Saved by Simon, one might say…

Stevin performed his experiment four-and-a-quarter centuries ago. In what follows, I will trace the various explanations that have been given for the dramatic fall that '*appear*[ed] *to be a single blow*' in the course of more than four centuries. In the process, we will pass a series of historical milestones that mark the road to the present state of physics.[7]

History does not end today, and Stevin's finding is still highly enigmatic. This is due to the discovery that lead is made of atoms, and that these atoms are made of yet smaller particles. I will sketch the relevant aspects of particle behaviour that are engraved on yet more milestones, beginning where the previous series ended.[8]

Having followed that road to the place where we stand today, we will see that two monumental achievements in theoretical physics, namely general relativity and quantum field theory, are in dramatic conflict with each other. This conflict may be cast in the form of the most important physics question of our time. For the moment, that question may be phrased as: *How does the Sun produce the curvature of its surrounding space-time?*

7 Kepler's *Harmonices Mundi* (1619), Galileo's *Dialogo* (1632) and *Discorsi* (1638), Huygens's relativity theory in *De Vi Centrifuga* (written in 1659, published posthumously in 1703) and *Horologium Oscillatorium* (1657), Newton's *Principia* (1687) and Einstein's *General Theory of Relativity* (1916).

8 Schrödinger (1926), Feynman (1948), Yang & Mills (1954), Englert & Brout (1964), and the recent discoveries at CERN's Large Hadron Collider (2012).

The Process of Progress

Do facts exist? Simon Stevin would have answered 'yes' without any hesitation. He did an enormous number and variety of experiments, including the crucial one on that tower in Delft: this is a historical fact. His experiments were quite repeatable, both during his time as well as today, and were often repeated and improved.[6] These are physical facts.

Do laws of nature exist? Not that we know of. Theories evolve, facts remain. Stevin's demonstration of the most remarkable property of falling bodies is as striking today as it was four centuries ago, even though in our time we see it demonstrated in the form of the so-called 'weightlessness' of astronauts in their spacecraft. I will follow the historical evolution of the concepts and theories related to Stevin's experiment. Along the way we see theories of motion, collision, accelerated motion, mechanisms that produce acceleration, gravity, space-time curvature, and the bizarre properties of matter in the form of quanta that are described by quantum field theory.

At the point when history becomes present, the path of this research bogs down in a marshy landscape where, at night, will-o'-the-wisps called 'supergravity' or 'string theory' spread a feeble and misleading light. I fervently hope that this book will inspire someone to find a way ahead. Arthur pulled a sword from a stone, helped by his tutor, Merlin. Maybe a 21st-century girl or boy will perform a comparable feat in physics, helped by a physics professor who teaches her or him that theory is the art of the possible.

When we follow the long and winding road from Stevin's beautiful experiment to present observations with giant telescopes and immense particle accelerators, we are confronted squarely with the evolution of scientific understanding: the same observation gives rise to an evolving sequence of explanations and theories. This demonstrates the provisional and temporary character of all results in physics. The phrase 'law of nature' is misleading, unless 'law' is meant to be similar to laws in society, which are made and amended as needed.

What is commonly called a 'natural law' is actually an intermediary link between the makeup of the Universe and our understanding thereof. There is no indication at all that this understanding converges little by little towards a single 'law'. The current theory of quantum electrodynamics is radically different from 19th-century theories of electromagnetism, which in turn differed enormously from Huygens's description of the propagation of light.

Often, older theories are still useful in their original context. For example, the propagation of water waves can be described perfectly well without explicitly taking into account that water is made of H_2O molecules. But when we wish to understand the bulk properties of water (its viscosity, density or wetness) we must dig deeper.

This is a Small Book about a Big Question, not a textbook of known physics. Therefore, it has to make do without the quantitative rendering of experiments and the mathematical description of theories. Much precision is lost thereby, because mathematics is miraculously useful in finding and formulating what is and what is not.

Stevin himself was an excellent mathematician, using math in the way a champion athlete uses oxygen: constantly, without fuss, and in large quantities. In his writings, he never seems to wonder why this works so well. But many of his successors have recorded their opinions about this unsolved mystery. For example, Einstein[9] wondered:

How is it possible that mathematics, which is a product of human thought independent of experience, is so admirably suited for tangible reality?

9 Albert Einstein (1879-1955), German physicist.

In a similar vein, Wigner[10] wrote:

> *The immense utility of mathematics in the sciences borders on the miraculous, there is no reasonable explanation for it.*

Brief and to the point is Galileo:

> *… just like something happens in reality, so also in the abstract…*

For what it is worth, let me add my personal note to this. I think that Einstein was prejudicing the issue by stating that human thought is independent of experience. In my opinion, this is demonstrably wrong. Our brain and all of its functions are products of biological evolution. It is a physical organ, like the heart or a kidney. It evolved for the usual biological purposes: survival and procreation. It turned out to be so useful that its purposes grew much more general. For example, monkey-like communication could become human language.

Just like our senses, our brain evolved to cope with 'experience', so that the probability of our survival increased. Therefore, I don't find it too audacious to maintain that our brain is arranged specifically to deal with 'tangible reality'. Diametrically opposite to what Einstein said, I think that it is no wonder at all that it is so 'admirably suited'. If it weren't, we'd be extinct, or a marginal species like most of the others. Of course, it doesn't follow from this that mathematics is a 'natural product'. But, given the great similarity between natural language and the language of mathematics, I venture that it probably is.

10 Eugene Paul Wigner (1902-1995), Hungarian physicist.

Laws Ain't

One of the beauties of Stevin's setup is that it uses a familiar, everyday effect. The result of the test triggers the most basic characteristic of the brilliant researcher. This isn't curiosity, as is commonly thought, but *perceptiveness*: the ability to see what everyone else can see, too, only better, or more connected to other things, more cleverly abstracted from circumstantial clutter, or more broadly generalized.

As an example, consider the statement *The sky is dark at night*. This truism is, on closer inspection, quite remarkable and non-obvious. Another example of a fact that is so strange that it borders on the bizarre: *I am not the average of my parents*, especially if we see it together with the fact *I resemble both of my parents*. The darkness at night is due to a subtle combination of effects.[11] The most prominent of these are: stars cannot emit more light than a certain maximum; the Universe expands; the Universe has a finite age. That people are not the average of their parents, even though they resemble both of them rather closely, turns out to be mostly due to the fact that we are built out of elementary particles.[12]

Everyday phenomena are almost completely incomprehensible in their raw appearance, so they are a very bad guide to physics. The world is messy, and there is not a single observation — whether in the days of Archimedes[13] or in the days of the European Extremely Large Telescope — that is free from the interference of side effects. To make matters worse, if we speak about physics, we are obliged to use words that have an established common meaning already, such as 'energy', 'force' or 'symmetry'.

It is useful to return one final time to the subject of 'laws of nature'. I will argue that these do not exist if, by 'law', we mean some sort of Ultimate Truth that will never change once we have discovered it. If

11 Edward Harrison, *Darkness at Night* (Harvard University Press, 1989).
12 Actually, I hesitate to use the words *particle* and *fundamental particle*. First, 'particle' suggests something like a marble or a billiard ball. But electrons and quarks behave in radically different ways. Second, we don't really know how 'fundamental' the known particles are. A better word would be *quantum*, for reasons given below.
13 Archimedes of Syracusa (287-212), Greek scientist.

science were a methodical process that zooms in on 'natural laws' by steps big and small, then we would expect that it would always be fairly clear what course to take. But the quest for the mechanisms of the Universe has led us to a point from which there is no visible road ahead.

The aura of absoluteness and infallibility of science is strengthened by the fact that scientific theories and results are cast in the language of mathematics.[14] Math is absolute, or so it would seem, because it is difficult to accept the possibility that there could be *two* versions of, say, Pythagoras's Theorem[15] (I do not mean 'two different proofs of the same theorem'). The numbers 3, 4 and 5 obey the equation 3×3+4×4=5×5. A plane triangle with sides of 3, 4 and 5 has a right angle at the corner where the sides of length 3 and length 4 come together. Any carpenter from the time of Aristarchos[16] knew this, and it doesn't seem to make sense that a craftsperson from a planet around the star Delta Orionis would hold a different opinion.

Moreover, mathematics starts with 'definitions' and 'axioms', statements that are taken as fixed and immutable starting points, from which theorems are derived; thus, math is quite similar to the ancient top-down philosophical 'world systems', and it is hardly surprising that many philosophers were and are well versed in mathematics. And yet math is curiously free, a product of invention. Just as in fiction or poetry, once you've thought something up, it exists. If you want to have a number with the property that its square is a negative number, then: hey, presto! It is there. Whether or not it is amusing or useful, is another story.

Following this top-down spirit of 'immutable laws' the most famous of all scientific theories, Newtonian mechanics, was built in a strictly mathematical fashion: first definitions and axioms, or 'principles', followed by logic and deduction. The idea that this is the proper approach was formulated earlier by Galileo, with his statement that in nature,

14 The scope of the language we call 'mathematics' is fiercely debated, especially after Gödel proved his famous *Incompleteness Theorem*, which states that any sufficiently rich mathematical system contains 'undecidable propositions'. That is to say: math contains true statements that cannot be proven to be correct, and false statements that cannot be shown to be wrong. He did this by constructing the mathematical equivalent of the undecidable statement 'All Cretans are liars, said a Cretan' or, even more brusquely, 'I am a liar'. See e.g. E. Nagel & J.R. Newman, *Gödel's Proof*, New York University Press, 1960. Kurt Friedrich Gödel (1906-1978) was a Czech mathematician.

15 Pythagoras of Samos (572-500), Greek mathematician.

16 Aristarchos of Samos (310-230), Greek astronomer.

things proceed as in mathematics.

The modern view of science, in which a scientific theory is not a law that is fixed forever but a temporary product of some kind of intellectual evolution, was formulated explicitly by Huygens on various occasions. Where Galileo held that 'As in Nature, so in mathematics', and where Newton and his successors took the view that 'Nature is mathematics, we only have to find its axioms', so Huygens's opinion may be paraphrased by 'As in Nature, so approximately in mathematics.' Huygens stated this view very explicitly in the preface to his book on the propagation of light:

One can see that these demonstrations do not offer as great a certainty as geometry, and may even occasionally differ strongly from it. This is because, while mathematicians prove their propositions by sure and incontestable principles, in this case, the principles are verified by reference to the conclusions drawn from them. The conditions of nature offer us no alternative. Still, it is possible to attain a degree of probability that quite often is hardly less than complete certainty. This occurs when those things that one has deduced from the supposed principles correspond perfectly to the phenomena that observations show us; especially when there are a large number of these, and still more powerfully when one formulates and predicts new phenomena that must result from the stated hypotheses, which then actually are found as foreseen. [17]

That point of view is directly opposed to the conviction that 'Laws of Nature' exist. In this respect, too, Huygens was far ahead of his time, and a leading innovator. [18] In the words of science historian Floris Cohen: [19]

17 Christiaan Huygens, *Traité de la lumière*, Pieter van der Aa, Leiden 1690.
18 The Persian mathematician Abu Ali al-Hasan ibn al-Haytham (called Alhazen in Europe, 965-1040) expressed similar views about the key role that experiments play in understanding the mechanisms of nature. He is sometimes called 'the world's first true scientist'.
19 Floris Cohen, *De herschepping van de wereld*, Bert Bakker, Amsterdam 2008.

But now, for the first time, an entire natural philosophy is treated by Huygens as a hypothesis, the utility of which is not assumed to begin with but must be tried and tested anew every time. By now, we are hardly aware that it was ever different; but before the years 1652-1656 even the possibility of such an approach was not considered.

A specific example of the process of progress is Huygens's explanation for the appearance of the planet Saturn. Contrary to common belief, Huygens did not *discover* Saturn's ring: he *explained* the planet's telescopic image. The discovery of Saturn's odd appearance was made by Galileo on July 25, 1610, when he wrote:

Saturn does not stand alone but is composed of three parts that almost touch each other; they do not move with respect to one another, nor do they change.

Even the best telescopes of the time, including those that were equipped with the state-of-the-art lenses that Huygens and his brother Constantijn[20] made, did not allow observers to see much more than what Galileo saw: a fuzzy planetary disk with a blurred protrusion on each side.

Dozens of explanations were brought forward by astronomers. Every one was a guess about the shape of the planet, or about the presence of two very fat satellites — planets, almost — flanking the main orb of Saturn. Of course, it was not known at the time that objects as massive and as large as Saturn must be almost perfectly spherical due to their own gravity.

Huygens's hypothesis about Saturn was inspired by his discovery of its brightest satellite, which he sighted on December 27, 1657 (later

20 Constantijn Huygens, Jr. (1628-1697), Dutch diplomat.

named *Titan* by John Herschel, son of the astronomer William[21]). He followed up his discovery and plotted the orbit, probably by using the micrometre arrangement for the eyepiece of his telescope, which he invented.[22] Assuming the orbit to be circular, he determined its inclination and period.

At that moment it must have occurred to him that the projection of Titan's orbit coincides with the direction of the 'handles' on the sides of the planet, suggesting that those protrusions were also in orbit. But because Galileo had clearly stated that the 'handles' do not change (an observation that had been well verified by others), the orbiting object could not be a single satellite: it had to be something that surrounds the planet on all sides.

Thus, I presume that Huygens assumed that Saturn did not only have the newly found satellite in orbit, but also a ring that lay in the same orbital plane as Titan did. He realized that a thin ring, when seen on its edge, would be practically invisible from Earth. In fact, observations had already shown that the planet's protrusions actually do change a little. Because Saturn's orbital period was known to be about 30 Earth years long, it was to be expected that once every 15 years, the ring should be invisible. Knowing the inclination of Titan's orbit, Huygens could — and did — predict exactly when the protrusions would be least visible.

Huygens had predicted that this would happen in May 1670. On the 27th of that month, he saw from the newly founded Observatoire de Paris, in the company of his colleagues Cassini and Picard,[23] that the planet's side lobes had indeed disappeared.

Huygens's book describing his interpretation of Saturn and its surroundings caused a big stir and fierce debate.[24] One of his opponents accused him of all sorts of bad things, including sacrilege and heresy, of course. Shortly after that, Prince Leopoldo of Tuscany asked Borelli,[25]

21 Friedrich Wilhelm Herschel (1738-1822), German astronomer; John Frederick William Herschel (1792-1871), English scientist.

22 The first known micrometre arrangement was constructed in 1639 by the young British amateur astronomer William Gascoigne. Huygens invented his micrometre independently, in his work on the improvement of the telescope eyepiece that still carries his name. See Henry C. King, *The History of the Telescope*, Dover Pub. 2003, pp. 95.

23 Giovanni Domenico Cassini (1625-1712), Italian astronomer; Jean Picard (1620-1682), French astronomer.

24 Christiaan Huygens, *Systema Saturnium*, Adriaan Vlacq, The Hague 1659.

25 Giovanni Alfonso Borelli (1608-1679), Italian scientist.

an Italian scientist, to decide the matter.[26] Borelli set up a model experiment. At a considerable distance from a group of observers (some of whom had never looked through a telescope before), a miniature model of Saturn with its ring was placed next to a candle. After observing this using the astronomical telescopes of the time, the group concluded that, indeed, the appearance of this setup properly reproduced what was seen in the sky.

Huygens's explanation of Saturn's appearance follows exactly the procedure he described in his *Traité de la Lumière*: Beginning first with observations (both his and Galileo's), which were viewed with perceptiveness (the projection of Titan's orbit lies in the same direction as the 'ears' on the planet), then following by forming a hypothesis (the ring) and, finally, ending with the verification of a prediction (the disappearance): '...and still more powerfully when one formulates and predicts new phenomena [...] which then actually are found as foreseen'[17].

There is no recipe for progress in physics, no method. Finding an explanation is not comparable to discovering an unknown island in the ocean, or a new animal species. The island and the animal exist already; all one needs to do is find them, with luck or spadework. But understanding a natural phenomenon is something that one has to do all by oneself: it is an act of creation.

Even though Descartes's[27] famous book bears the title *Discours de la méthode pour bien conduire sa raison* (*Discourse on the Method of Rightly Conducting the Reason*), it is not a cookbook with recipes that lead to a Nobel Prize. It could not be, because discoveries are the occasional product of research. Occasional, because the primary product of research is failure. Not to worry: in a sense, even a 'failed' experiment or theory helps, if only by showing what direction we should not take. In the words of Huygens: *the conditions of nature give us no alternative.*

26 See Luciano Boschiero, *Experiment and Natural Philosophy in Seventeenth-Century Tuscany*, Springer 2007, ch. 8.

27 René Descartes (1596-1650), French scientist: *Discours de la méthode pour bien conduire sa raison et chercher la vérité dans les sciences*, Jan Maire, Leiden 1637.

Motion

With five centuries of hindsight, it seems that the development of mechanical theories proceeded predictably. In the year 1600, however, it was by no means clear where to begin, what to study or what questions to ask. One of the key items on the physics agenda was *change*. The ancient Greek philosophical dictum, *panta rhei* ('everything flows')[28], lacked precision and was unacceptable in *experimental* philosophy as an explanation for change.

Change presents itself in many ways, the most obvious being the motion of objects. In 1600, common sense seemed to show that motion needs an agent to produce and to maintain it, and everything seemed to move towards the ground, unless something interfered. The first person in the history of our planet to make serious headway on this subject was Galileo. He did this by replacing the philosophical question, 'What is motion?' with the experimental-philosophical inquiry, 'How does motion behave?'

With his immense perception and talent for precise and systematic experimentation, Galileo began to study the behaviour of spherical balls that were set up to roll down inclined planes so that they moved more slowly than simply dropping vertically. In a long series of experiments, he discovered most of the basics of falling motion.

First: the velocity increases in direct proportion to time (in the absence of perturbations such as the resistance of the air). That is to say: free fall is uniformly accelerated motion. Second: the speed that an object acquires when released is always the same after it has fallen a given vertical distance, in free fall as well as when constrained to move on an inclined plane.[29] Third: Galileo deduced and verified that it follows from the first finding (speed is proportional to time) that the distance an

28 The aphorism *panta rhei* was the central theme in the philosophy of Herakleitos of Ephesos (540-480).
29 This is not exactly correct for balls that have a finite size. In modern terms, we ascribe this to the fact that the rotation of the ball represents a certain amount of energy. The ball picks up this energy at the expense of its forward speed during its rolling down the plane.

object traverses when falling is proportional to the square of the time. Fourth: horizontal and vertical motions occur independently. From this, he concluded that the path of a thrown mass is a parabola.

In parallel with these experiments, Galileo developed a view of the more general and abstract properties of motion. The consensus at the time was that motion requires something to keep it going, with the exception of motion 'above the Moon', that is, the motions of the planets. These were supposed to be a composite of various forms of uniform circular motion.

Instead, Galileo supposed that it should be possible for motion to maintain itself indefinitely without recourse to an extra agent. The question then was: what sort of motion could go on forever?

In his *Dialogo*, Galileo tackles this question in his characteristic mathematics-driven way.[30] Through his alter ego Salviati, he states that a circle does not change when it is rotated about its centre. Therefore, he argues, circular motion can maintain itself indefinitely without having to be driven. Next, he says that a straight line does not change when it is displaced along itself, either. Thus, rectilinear motion is also a candidate for being 'the' motion that can persist indefinitely. Finally, he makes contact with the physical world. Geometry says that a straight line is infinitely long, but: 'we know that the Universe is not infinitely large'. Therefore, straight-line motion is excluded and only circular motion remains as the true 'natural' motion.

In so doing, Galileo singled out circular motion in an entirely logical way, demonstrating that hindsight-driven disdain for 'epicycles' is not justified. In fact, his line of argument sounds strangely modern: deriving a 'law of motion' from a symmetry principle is probably the greatest success story in theoretical physics (I will discuss symmetries below in the section *A Twist to the Tale*).

30 Galileo Galilei, *Dialogo sopra i due massimi sistemi del mondo*, Landini, Florence 1632, *Dialogo Primo*, p. 43.

But Galileo's argument is wrong for various reasons. The most important one is that the symmetry of his circle is broken. A planet is a dot in space, not a curve. Even if its orbit were circular, with the Sun at the centre, a planet is not like the solid rim of a perfectly uniform wheel with the Sun at the position of the axle. A planet is a very non-circular object moving in space, rather like the tube valve on a wheel, which drastically changes the situation. A photograph with a very long exposure time would show the planet smeared out along its orbit, but a snapshot catches it at one specific location. In modern terms, we would say that planetary motion is a case of 'spontaneously broken symmetry'.

Huygens's Relativity

Christiaan Huygens was the first to prove explicitly that Galileo's argument about 'natural motion' is wrong. He replaced this idea through a number of steps. First, he postulated a 'principle of relativity' that he supposed to be valid for all motions. He did not introduce this item as an axiom, like a mathematician or a classical philosopher would, but as a summary of what he perceived as the most striking characteristic of motion — without, of course, including some sort of 'natural motion' from the start.

From that, he deduced what later came to be called the 'law of inertia': free or 'natural' motion is motion with a constant velocity,[31] that is, motion in a straight line with constant speed, flatly contradicting Galileo's assertion[30] about circular motion. Next, he computed geometrically what the difference is between Galileo's two forms of 'ideal' motion: the circle and the straight line. In the process he derived the first-ever algebraic equations in theoretical physics, describing the centrifugal acceleration and the oscillation time of the ideal pendulum.

Huygens's main conclusion forms the next step on the path leading from Stevin to modern physics: *a curved orbit is an accelerated orbit*. An object maintains a constant velocity (fixed speed in a fixed direction) with respect to other objects, unless — through some outside agent to be specified — an acceleration interferes.

But let me start at the beginning. As I have argued above, the primary characteristic of a great scientist is not curiosity, but perceptiveness: the ability to see what others have also seen, but from a different angle. The vision of 'what motion is' was vaguely present in some of the writings of Galileo and Descartes,[32] but Huygens provided the clear and definitive formulation. It occurred to him that motion does not

31 In physics, 'velocity' is not the same as 'speed'. *Speed* is the distance covered in a given amount of time, in whatever direction. *Velocity* is speed in a specific direction. Technically, speed is a *scalar* (just a number, such as 300,000 km/sec), whereas velocity is a *vector* (usually symbolized by an arrow, the pointing of which indicates the specific direction and the length the speed).

32 With some hesitation, I have decided to leave Descartes' contributions to this story aside, because his view of dynamics is too tied up with his 'vortex theory' of forces. His atomic-based interpretations (to call these 'theories' would be too generous) of cosmic motion postulate all sorts of non-quantitative properties, and his successors dreamed up myriads of particle theories constrained only by their imaginations.

mean that an object changes its position in space, but that its position changes relative to other objects in the Universe. The sentence he wrote in his notes reads:[33]

Motus inter corpora relativus tantum est.
Movement between objects is relative in all aspects.

That is two symmetry principles rolled into one. First: nothing changes if you change all positions in space by shifting all positions by the same amount. Second: nothing changes if you add a fixed velocity to all velocities throughout space.

The physics of this statement is as plain as it is remarkable: position and velocity are not intrinsic properties of objects.[34] You don't know where you are, and you don't know how fast you are going where. Position and velocity can only be observed with respect to other objects. This is a typical case of everyday experience contradicting physics. If I were to ask, 'Conductor, does the Cambridge railway station pass by this train?' he or she may think I'm strange, but actually I am just being a physicist.

Huygens's principle of relativity pulls the rug out from under the question 'What is it that moves?' If the Universe were to contain but a single particle, then it would be impossible to state whether it moves or not. In other words, apparently our Universe is built in such a way that there is no meaningful distinction between 'rest' and 'steady motion'. In his writings, Huygens is quite explicit:[35]

We observe that it is impossible to determine whether objects are at rest or in motion, unless with respect to other objects. [...] In vain would one ask what that true motion is, and what is the use of that anyway?

33 Huygens's manuscripts, book 7A, folio 24 recto, University Library, Leiden. This was probably written around 1688, but he wrote similar texts (albeit less pithy than these six words) much earlier.

34 Today we know that time is relative, too: only time differences are observable; Huygens, however, made no statements about the properties of time.

35 Christiaan Huygens, *Oeuvres Complètes* XVI, pp. 111, 228. Idem, letter to Henry Oldenburg, August 10, 1669.

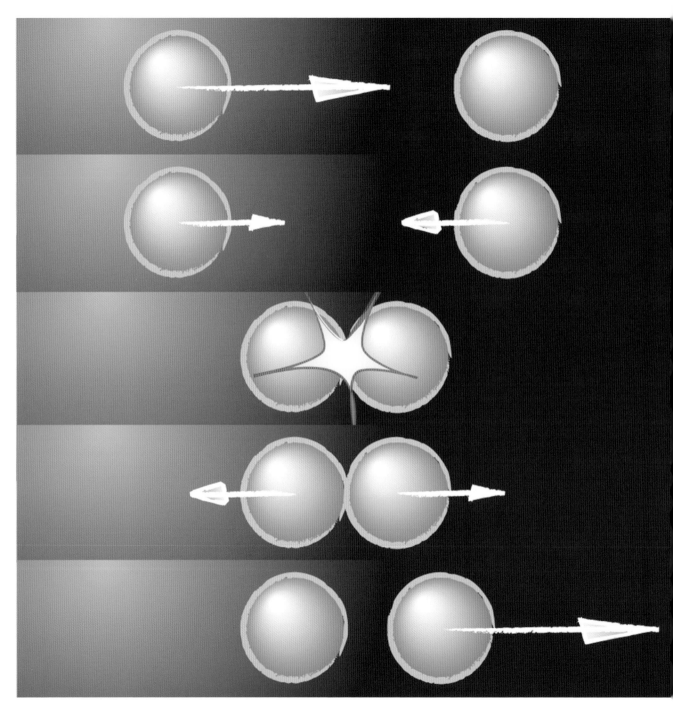

... in my opinion, motion and rest can only be relative, and the same object that some say to be at rest, may be said to move with respect to other objects, and therefore one kind of motion is no more true than another.

In Huygens's relative world, position has no meaning; only differences of position count. In the 17th century, ships navigated using different longitude systems; a Dutch merchant would use maps based on the Amsterdam meridian, whereas a French vessel would use Paris. No matter, because the distances travelled on Earth are all differences, and a rotation of 2.54 degrees around the Earth's axis would make the maps coincide. Likewise, says Huygens, if we were to take the whole Universe and move everything in it over an arbitrary but fixed distance, after this

← Relativity of motion in a system of two identical spherical balls in a central collision. Strip 1: An observer at rest with respect to the ball on the right sees the left one move with speed *v* from left to right. Note that another observer, standing still with respect to the ball on the left, would see the right one moving with speed —*v* from right to left. Strip 2: When the observer changes his/her 'frame of reference' by moving to the right with half the speed *v*, he/she sees the ball on the right moving leftward with speed -*v*/2, and the ball on the left moving to the right with speed *v*/2. Strips 3/4: The collision reverses the directions but not the values of the relative speed. Strip 5: When the observer returns to his/her original 'frame of reference', the incoming ball is at rest and the other moves with speed *v* to the right. The relative speed of one ball with respect to the other is always *v*, except at the very instant of the collision. What happens there depends on the internal constitution of the balls, which is extremely complex.

operation everything would be exactly as before.

This is what physicists call a *symmetry*. It is very closely related to Galileo's arguments about 'natural motion': *changing something so that nothing changes*. A difference in position, divided by the time it took to effect that difference, is a velocity. But velocities are also relative. If we were to take the whole Universe and give everything in it an extra velocity, the same everywhere but otherwise arbitrary, the whole world would run exactly as before this operation.

It follows that an observer who moves with constant velocity (fixed speed in a fixed direction, often called 'inertial motion') cannot determine whether she or he is moving or not, other than by referring to surrounding objects. It is then pointless to ask whether a specific constant-velocity motion will 'persist' or not. Because our Universe does not allow absolute positions or velocities, inertial motion is the 'natural' motion that Galileo sought.

The depth and power of Huygens's relativity is evident in many ways. One: 'Newton's first law' (in the absence of forces, objects move with constant velocity) is a consequence, as we saw above. Two: it leads to the formulation of collision laws, in which the central role of mass becomes evident. Three: it focuses attention on that which is *not* relative, namely *acceleration*.

To get some idea of how this works, consider two identical spherical balls moving towards each other on a line through their centres. What happens after the collision may be seen by using the relativity of motion throughout. We shift our point of view twice, but 'Huygens's relativity' implies that the process we are observing does not change.

Suppose that an observer sees a ball on his left moving with speed *v* towards an identical ball on the right, which is standing still as seen by the observer.[36] The balls collide. What will their state of motion be after the impact? By strictly using Huygens's relativity, we find an answer (and we can see how useful math is, by the way — in this case, just addition and subtraction).

In what follows, a positive speed means that the ball moves from left to right, whereas a negative speed has the ball moving from right to left. Seen from this point of view (the technical term is *frame of reference*), their relative velocity is $v - 0 = v$. The motion of the balls, as seen in this frame, is not symmetric: one moving with v, the other has speed 0. But the view can be made symmetric if the observer changes the frame of reference, by moving with constant speed of one-half v (that is $v/2$ from left to right). That means that *all speeds are reduced* by an amount $v/2$. According to Huygens, velocity is entirely relative, so this will not be altered by the collision.

In the new frame, the left ball moves with speed $v - v/2 = v/2$, the one on the right with $0 - v/2 = -v/2$, so that their motion is symmetric. Their relative speed is still $v/2 - (-v/2) = v/2 + v/2 = v$. In the new frame of reference, where the speeds are equal and opposite, the speeds are swapped by the collision, which does not change the symmetry. Their velocities are then $-v/2$ and $+v/2$.

Now the observer goes back to the original point of view, by *increasing* all speeds by an amount $v/2$. In that frame, the speed of the balls after the collision will be $-v/2 + v/2 = 0$ on the left and $v/2 + v/2 = v$ on the right.

After the collision, the incoming ball stands still with respect to the observer, and the other moves with speed v. Is this what happens? A simple experiment on a billiards table shows that it is. If one ball is lying still with respect to the table, and another identical one moves towards it and hits it dead centre, the incoming ball stops and the target ball moves away with the speed of the incoming ball.

36 It may seem a bit pedantic and tiresome to keep inserting phrases like *as seen by the observer*, or *in the observer's frame of reference*, but this is essential in classical relativity: there are no preferred positions, directions or velocities in our Universe.

Huygens proceeded by considering what happens when a ball hits two identical balls glued together. In that case, it is not permitted to simply swap them, because the situation is not symmetric. However, Huygens knew what to do: he incorporated the detailed investigations done by Stevin on the equilibrium of the lever. Place a ball at one end of a weightless beam and two such balls at the other end. Then the beam is in equilibrium if it is supported on a pivot that is two-thirds of the beam's length away from the single ball (so that it is one-third of the beam's length away from the pair).

If the observer moves with speed $v/3$ to the right, he sees that the balls collide exactly at this pivot, which does not move with respect to the observer. Because of Stevin's argument about the lever beam, which is always in equilibrium with respect to that pivot, the single ball can be swapped with the pair. The single ball then has a speed $-2v/3$, and the two balls $v/3$ each. Going back to the original state of motion of the observer, the single ball is seen to move with $-2v/3 + v/3 = -v/3$, while the two balls have $v/3 + v/3 = 2v/3$. Notice that the relative speed

$\rightarrow$ Collision of a spherical ball with two other identical balls, glued together (strip 1). If the balls were at rest and connected by a weightless rod (strip 2, yellow line), the whole assembly would be in balance if supported at a pivot two-thirds along the length of the rod (orange dot, strip 2). Today we call that point the *centre of mass*. By considering the motion with respect to that point, Huygens could deduce the collision law for un-equal masses. Let the incoming object have the speed v as seen by an observer, hitting a target that has twice the mass of the incoming ball and stands still with respect to that observer. The incoming ball is seen to recoil with speed $-v/3$, while the target is propelled to $+2v/3$ (strip 4). Their relative speed is still $2v/3 - (-v/3) = 2v/3+v/3=v$ after the collision.

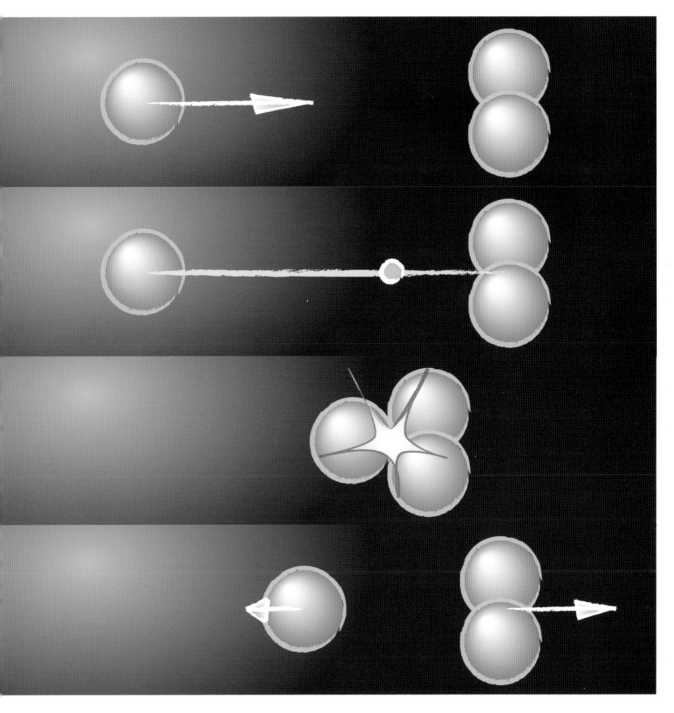

of the single ball with respect to the pair after the collision is $2v/3 - (-v/3) = v$, as it was before the collision.

Again, experiment shows that this is what happens. Nobody would have arrived at this curious result by just guessing: one ball bounces back with speed $-v/3$, the two balls are propelled to a speed $+2v/3$ each. In fact, the person who did such guesswork, Descartes, was wrong on most counts. Huygens flatly contradicted the statements made by the famous philosopher, to the dismay of Van Schooten,[37] his Cartesian mathematics professor.

37 Frans van Schooten, Jr. (1615-1660), Dutch mathematician.

Acceleration

If I ask the train personnel, 'Conductor, does the Cambridge railway station pass by this train?' I'm being a bit strange, but not wrong, because velocities are relative. However, saying, 'Conductor, does the Cambridge railway station stop at this train?' makes no sense in our Universe.

Because position is relative, we must use the *change* of position in order to describe motion. This change we call *velocity*. But velocity is relative as well; therefore we must use the *change* of velocity in order to describe motion. This change we call *acceleration*. This is, in fact, an observable, as every cyclist knows who has had to brake for a traffic light.

The word is a bit specialized, because in physics we use the word 'acceleration' for every kind of change in velocity: an increase or a decrease of speed as well as a change in direction are all called by this name. Velocity has a direction and a magnitude,[38] so that a change of direction counts as an acceleration too, even if the speed (number of metres travelled per second along a path) remains the same.

That is precisely the case with Galileo's circular motion. The speed remains constant, but the direction changes steadily. The question then is: how does that feel? In a brilliant sequence of arguments, Huygens equated the acceleration of uniform circular motion to the steady acceleration of a falling object. Thus, he could relate the pull which is felt on the string of a slingshot directly to the acceleration of gravity.

From Galileo's work,[39] Huygens knew that the speed of a falling object increases linearly with time. If the amount of acceleration is arbitrarily set to 1, and an object starts at speed zero, then in 2 units of time it reaches speed 2. The mean speed in that interval is then (0+2)/2=1. The mean speed in the next interval is 1+2=3, so that the mean velocity follows the sequence of odd numbers: 1, 3, 5, 7... at successive instants of time. The

38 That is why a velocity is graphically rendered as an arrow, the size of which indicates the magnitude (speed) and the orientation the direction of the velocity.

39 Galileo published his treatment of free fall in *Discorsi e dimostrazioni matematiche intorno a due nuove scienze*, Elsevier, Leiden 1638. See p. 173 ff. of the English translation by H. Crew & A. de Salvio, Dover, New York 1954.

distances travelled at these times are then the sums of the numbers: 1, 1+3, 1+3+5,... which add up to 1, 4, 9, 16... These are all *square* numbers: 1×1, 2×2, 3×3, 4×4..., from which it follows that the distance travelled by a falling object increases quadratically with time. From that, Galileo had proven that the orbit of a falling object is a parabola.

Huygens was a master geometer. He took Galileo's parabola and fitted it snugly around a circle.[40] This gave him the connection between the acceleration of a falling body and the acceleration that is needed to keep an object moving on a circle. His mathematical expression for the difference between linear and circular motion with constant speed led him to the first-ever exact equation in theoretical physics: $g = v^2/R$. In words: the amount of acceleration g needed to keep an object moving with constant

→ Constant acceleration means that the velocity changes by equal amounts in equal time steps. If an object is dropped from rest with constant gravitational acceleration g, then the speed V with which it falls is equal to the product of g and the time T, so that $V=gT$. The distance an object travels is equal to its speed multiplied by the time during which it has moved. If the speed changes (as it does when accelerated), the *added* distance it travels due to the acceleration is equal to the average of the *extra* speed and the time during which the acceleration has acted. Therefore, the height H of the falling object decreases by an amount equal to the area of the white triangle: one-half times the product of the extra speed v multiplied with the time step t. Because we already had $V=gT$, this means that an object starting from rest at height zero has fallen to a depth $H= -gT^2/2$ during the time T: the distance fallen increases with the square of the time. The shape of this curve is a parabola in space-time. If the object falls while moving forward, the resulting orbit is a parabola in space.

40 The technical term is: fitted with a second-order tangent. This happens when the focal distance of the parabola is equal to half the radius of the circle.

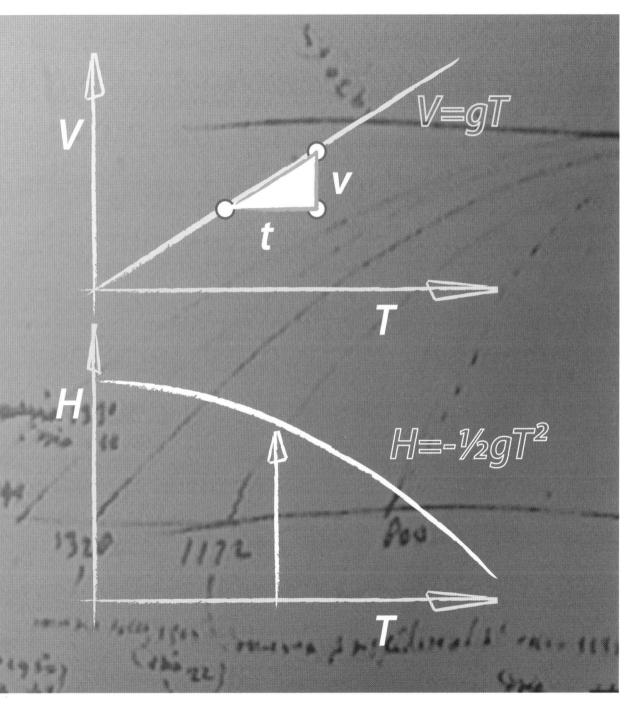

CHRISTIANUS HUGENIUS

DE

VI CENTRIFUGA.

GRAVITAS est conatus descendendi. Ponendo itaque gravia Acadentia Dve ad perpendiculum fue... velocitatis motu ea acceleratione, ut ... oribus æqualibus æqualia accrescant celeritatis momenta, certissime inde demonstrari potest spatia diversis temporibus e quiete peracta esse inter se, sicut temporum quadrata. Hoc autem experientiæ exacte convenit. Ergo recte illud assumptum esse constat. Exacte convenire experimenta Galilei, Riccioli, nostra comprobant, nisi quod acri resistentia pauxillum quid aberrare faciat, sed hoc eo minus, quo corpora plus gravitatis pro superficiei magnitudine continent, quoque in minoribus spatiis peri-

$$g = \frac{v^2}{R}$$

speed v on a circle with radius R is equal to the square of the speed, divided by the radius of the circle. The direction of the acceleration is towards the centre of the circle (and therefore perpendicular to the velocity).

Huygens derived the formula $g = v^2/R$ for the acceleration g that is required to keep an object moving on a circle. Because the acceleration is directed towards the axis of rotation, it is actually a *centripetal acceleration*: an acceleration 'seeking the centre'. However, everyone knows this effect under the name of *centrifugal force*, even though that term is doubly wrong. First, because it is not a force but an acceleration, that is to say, a change in velocity. Second, because the acceleration makes the object deviate from its straight constant-velocity path; it does not drive the object away from the circle, but does precisely the opposite.

Actually, Huygens himself used the expression *vi centrifuga*. Countless useful pieces of equipment use it, but the name is quite wrong. The word implies that an object moving in a circle 'flees' (*fuga*) from the centre. Quite the opposite is true: the object stays on the circle only if it is forced to do so. Centrifugal force is not a force at all, it is a tangible consequence of the relativity of motion.

← Galileo had shown that the trajectory of a freely falling object is a parabola, if it is given an initial forward velocity (when dropping straight down the path is a plumb line). Huygens knew that a circle fits snugly in the hollow of a parabola if the circle has a radius equal to twice the focal distance of the parabola (top image). Thus, he could relate the constant acceleration of gravity to the acceleration g that is required to make an object with a velocity v move on a circle with radius R. Instead of following a constant-velocity path (bottom left, white line) the object 'drops' onto the circle (blue line). In the crucial step, Huygens used his beloved geometry (bottom right) to show that the acceleration g equals v^2/R.[41]

41 Thus, the mathematical expressions for classical mechanics are not simply algebraic, such as $y = x^2$. Instead, they are prescriptions for 'the change of the change of position'. The technical term is *second-order differential equation*.

No force is required to squeeze the water out of your laundry in the spin dryer. To the contrary, force is required to keep the fabric in the spinning cylinder. That restraint is produced by the stress in the metal. In places where the metal is perforated, no restraining force is present and the drops of water race straight ahead, on a path that is tangential to the cylinder — and not radially outward, as many people think.

Classical mechanics provides an answer to the question, 'Where is what when?' in a doubly indirect way.[42] Position and velocity are relative; therefore, only the change of velocity — acceleration — plays a key role. An acceleration is 'the change of a change'. Because of the relativity of position and velocity (and time), the initial values of the position and velocity of all objects must be specified, together with a prescription for their accelerations. With that, the system is determined, and the whole future of the system is fixed by that initial point in the past.

However, the emphasis in mechanics is habitually not placed on the acceleration, but on something different, namely, a force. Force and acceleration are proportional to each other. The constant of proportionality is called *mass*. The acceleration of an object with a given mass is equal to the force acting on that object, divided by its mass.

Note that this is, in fact, a definition of mass, once we specify what a force is. By turning the equation around, we see that a force equals acceleration multiplied by mass. Thus, a given force causes a small mass to be accelerated a lot, while a big mass is accelerated only a little by the same force. Therefore, mass is a measure of the 'inertia' of an object, its 'resistance to change of velocity'.

If the force is proportional to an object's mass, we find something remarkable. In that case, the force is large when it acts on a big mass and small when it acts on a small mass. The net result is that every object

42 Here's the complete argument. The triangle CAB, spanned by half the distance AE travelled along the circle, has the same shape as AED, where AD is the distance the object would have travelled without the acceleration towards the centre C, and DE is the distance the object has 'dropped' due to the acceleration g in a small amount of time t. Thus, the ratio AB/AC is equal to DE/AE. Now AC is equal to the radius R of the circle. The length AD equals vt, DE equals $gt^2/2$ according to Galileo, and AB equals $vt/2$. Actually, AD is not exactly equal to AE, but the difference vanishes when the time step t is taken very small so that both D and E approach A arbitrarily closely. Now putting in these various quantities, we find that $vt/2R = gt^2/vt = gt/2v$. The factors 2 cancel, as do the factors containing t, and we end up with $g = v^2/R$.

that responds to this special type of force experiences the same acceleration, independent of its mass — which is what Stevin's experiment on the Grote Kerk in Delft showed, back in 1585. We will encounter this again when Einstein enters the story, just over four centuries later.

Gravity

In classical mechanics, the acceleration of objects has to be prescribed before the resulting motions may be computed. In fact, the mathematical formula that specifies how objects respond to forces, the 'equation of motion', can be read in this way: *The acceleration experienced by any object is equal to the net force exerted on that object, divided by its mass.*

The cause of the acceleration is not part of the system proper. It must be specified separately, put in from the outside, so to speak. When the science of mechanics was developed, this requirement led to a wild variety of hypotheses about the properties and causes of accelerations, to the tune of fierce and often acrimonious debate. In fact, the whole concept of 'force' had a bad reputation. It was much too vague, and carried the odium of magic and arbitrariness.

Of course every blacksmith, carpenter and bricklayer of that time knew that the forces among material objects were somehow related to their structure. The opinions of craftspeople and engineers, however, were not held in high esteem, except by people like Stevin and Huygens, who were very skilled in engineering and practical work.

Nor did it help that Descartes made a fine mess of it with his vortices of hypothetical ghost particles. Armchair philosophers could, and usually did, invent a new particle for every phenomenon in the world, resulting in a Shakespearean 'sound and fury, signifying nothing'.

So when Newton declared that 'gravity' was the 'universal' source of accelerations, he did something quite daring. To his and our good fortune, he also cast the expression for his 'universal force' in an astonishingly simple mathematical form: the acceleration is inversely proportional to the square of the distance and independent of the accelerated object's mass. The latter requirement was absolutely necessary because of Stevin's experiment in 1585, about a century before Newton's work.

Scientists who are sensitive to such things find this simplicity a source of great beauty. It seems instantly convincing because it is the diametrical opposite of all the vague Cartesian stuff about ethereal particles and vortices, which contains no more physics than the 'epicycles' of antiquity.

By way of illustration, consider the combination of Huygens's formula for the centrifugal acceleration with Newton's gravity. Huygens wrote $g = v^2/R$, and according to Newton, this was $g = 1/R^2$. Combining these two expressions, we conclude that $v^2 = 1/R$: the square of the orbital speed is inversely proportional to the radius of the orbit.

This is precisely what Kepler[43] had concluded from the astronomical observations he inherited from Brahe.[44] Kepler published this in *Harmonices Mundi*, and it has been called Kepler's Third Law ever since.[45] In fact, both Newton and Hooke independently derived the expression for the gravitational acceleration.[46] They applied the above argument in reverse: knowing Kepler's Third Law and Huygens's $g = v^2/R$, it follows that gravity obeys $g = 1/R^2$. Thereafter, they quarreled bitterly over who was the first to discover this.

Discoveries such as these established Newtonian mechanics at the top of 'experimental philosophy'. But under the surface grave problems remained. Maybe the thickness needed for the bricklayer's walls could now be computed, but what happened inside the bricks? Soon it became clear that gravity cannot make stable objects because its force is purely attractive.[47] Then what does matter have to do with gravity? That is a question which will come back in a number of guises, ultimately leading to the biggest question for the 21st century.

Meanwhile, Newtonian mechanics was long thought to be a Theory of Everything. It even had social and moral implications. These were origi-

43 Johannes Kepler (1571-1630), German astronomer.

44 Tycho Brahe (1546-1601), Danish astronomer.

45 Johannes Kepler, *Harmonices Mundi*, Io. Plancus 1619. Kepler gave this expression in a different but equivalent form: the square of the orbital period is proportional to the cube of the size of the orbit.

46 Robert Hooke (1635-1703), English physicist.

47 This emerged long after Newton, and the proof is fairly subtle. A system of two bodies in Keplerian orbits is stable. In a many-body system, a small fraction of the bodies forms a compact cluster, casting the rest out to infinity.

nally formulated by Leibniz,[48] who was the first to construct a computing machine that could perform all four arithmetic operations on whole numbers: addition, subtraction, multiplication and division. Just like La Mettrie[49] later did, he considered our brain to be a super-miniature computing clockwork, so he wrote about an ideal world in which agreement reached by means of reason and calculation would prevail over dissent based on mere opinion or conviction:

Whenever a conflict arises between philosophers, they need not put more effort into their scientific discourse than two professional calculators would. It will be sufficient to take up pen and paper, go and sit before the computing machine and say to one another (in a pleasant way, if possible): Let us calculate.

48 Gottfried Wilhelm Leibniz (1646-1716), German mathematician and philosopher: *De scientia universali seu calculo philosophico*, 1680.
49 Julien Offray de la Mettrie (1709-1751), French surgeon and philosopher: *l'Homme machine*, Leiden 1748.

Absoluteness Theory

Classical mechanics is a true theory of relativity. *Motus inter corpora relativus tantum est*; position and velocity are not properties of an object, only relative positions and velocities are observable. The equations of motion, called 'second-order differential equations', are the expression of this observation. It follows, too, that constant-velocity motion is the 'ideal', 'natural' or 'inertial' state of motion.

In Huygens's relativity, it makes no difference whether one is moving with constant velocity or standing still: according to the *motus* line, there is no way to decide between the two. The mere existence of an object is indistinguishable from its moving with constant velocity.

Unless, that is, our Universe has the property that some objects in it cannot stand still. In our everyday world, we are never aware of objects that cannot stand still with respect to us. It may take some effort, but it is perfectly possible to fly alongside a jet plane such that the velocity difference between you and the jet is zero.

But in a very surprising experiment in 1887, Michelson and Morley discovered that there is, in fact, something in our Universe that cannot stand still: light.[50] Up to that moment, it was thought that light — of which Huygens had successfully argued that it is a shock-wave phenomenon — must move with respect to some carrier medium, like waves on the surface of water are moving with respect to the underlying water volume. To their immense surprise, Michelson and Morley found that light does not behave like that. The speed of light — traditionally called '*c*' — is *invariant*. That is to say, light always moves with the same speed, no matter what the speed of the emitter or the receiver of the light is. Light rays cannot stand still with respect to anything. Huygens's assertion that *the same object that some say to be at rest, may*

50 Albert Abraham Michelson (1852-1931), Polish physicist, and Edward Williams Morley (1838-1923), American physicist: 'On the Relative Motion of the Earth and the Luminiferous Ether', 1887, *American Journal of Science*, 3rd series, pp. 34, 333.

be said to move with respect to other objects turned out to be false after all, at least for light.

A little thought shows that, therefore, all light moves with the same speed. In summary: the speed of light is not relative, but absolute. Had he known this, Huygens would have written *Motus inter corpora relativus tantum est, praeter lumen* — movement between objects is relative in all aspects, with the exception of light.

This finding was so massively odd, that unconventional explanations were invented. For example, Lorentz and FitzGerald[51] independently proposed that moving objects become foreshortened in their direction of motion. This fixed some aspects of the problem, but why this 'Lorentz-FitzGerald contraction' occurs, they could not say. Even though the physics did not turn out to be fruitful, the proposal attracted some attention:

A fencing instructor named Fisk
In duels was terribly brisk.
So fast was his action
The FitzGerald contraction
Foreshortened his foil to a disk.[52]

Einstein took the invariance of the speed of light seriously.[53] This implied that all of classical mechanics had to be rewritten. Einstein did that in his Special Theory of Relativity. This is, in fact, Huygens's theory of relative motion with constant velocities, but taking account of the fact that the speed of light is always the same (we will see below that the 'general' theory is the 'special' theory in which accelerations are included).

Among many other discoveries, he showed that three things follow immediately from the invariance of the speed of light. First: speed is a

51 Hendrik Antoon Lorentz (1853-1928), Dutch physicist; George Francis FitzGerald (1851-1901), Irish physicist.

52 Quoted in W.S. Baring-Gould, *The Lure of the Limerick*, 1970 Panther Books, p. 19.

53 Historians of science differ on this point. Einstein never clearly stated that he knew of the Michelson and Morley experiment, but he did not explicitly deny it either. Technically speaking, he could have deduced the invariance of *c* (the speed of light) from Maxwell's electromagnetic theory. From Maxwell's equations, it can be simply shown that all electromagnetic waves propagate with the same speed, namely *c*.

distance in space divided by an amount of time (in the convention of the *Système International*, the SI units, this is metres per second). The existence of a speed that is always the same implies that space and time can be measured with the same unit, namely the second. The distance to the Moon is 1.3 seconds, to the Sun 8.3 minutes, to the Andromeda Galaxy it is two million years. This unification means that space and time can henceforth be considered as a single structure, called (3+1)-dimensional 'space-time'.

Second: because the speed of light is absolute, time and space are relative. It is quite easy to demonstrate this by means of an idealized 'Lorentz clock': two parallel mirrors with a light ray bouncing up and down between them. Every reflection marks a point in space-time (called an *event*), and an observer may determine the time intervals between successive events. The time between reflections is equal to the distance between the mirrors divided by the speed of light. For example, if the mirrors are 300 metres apart, an observer who is standing next to the clock measures one millionth of a second between events.

Now let this clock move with respect to the observer, in a direction parallel to the mirror planes. With respect to this observer, the light follows a sawtooth path between reflections. As before, the time between reflections is equal to the distance traveled between the reflections, divided by the speed of light. In this case, however, the distance between events is *greater* (due to the diagonal motion of the light with respect to the observer) while the speed of light is still *the same*. The conclusion is that a moving clock is seen to tick more slowly than the same clock at rest with respect to the observer. This is called *time dilation*, and Einstein's relativity derives its name from it: because the speed of light is absolute, time is relative.[54]

54 By application of Pythagoras's Theorem to the triangle formed by the perpendicular light path (orange), the distance covered by the upper mirror (white), and the diagonal light path (yellow), one may easily show the following. If t is the time kept by a clock standing still with respect to the observer, and T is the time indicated by the same clock moving with speed v relative to that observer, then the length of the white path is vT, the length of the orange path is ct, and the length of the yellow diagonal is cT. Application of Pythagoras's Theorem to the white-orange-yellow triangle gives the result $T = t/\sqrt{(1-v^2/c^2)}$. Incidentally, the factor $\sqrt{(1-v^2/c^2)}$ is exactly the same as was proposed by Lorentz and FitzGerald when they postulated that moving objects are seen to contract in their direction of motion.

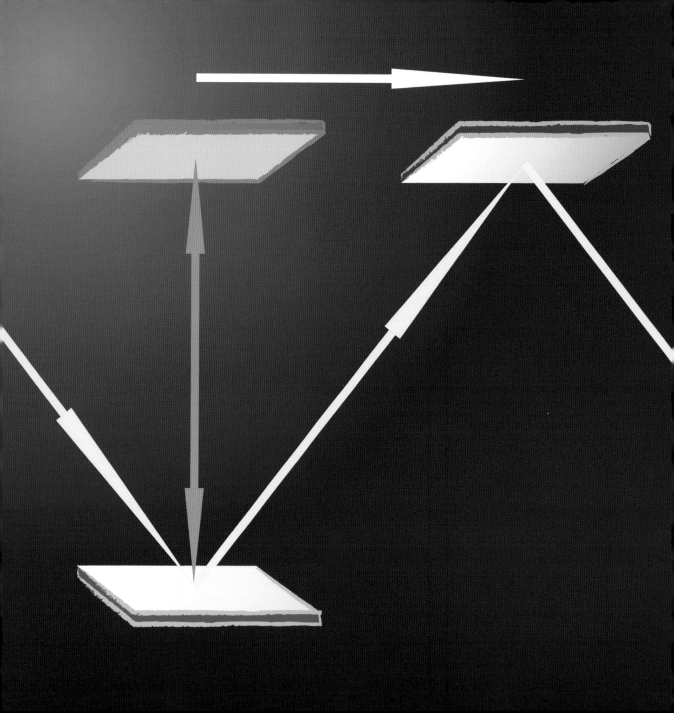

In everyday life, it seems that we can synchronize clocks whether they move or not: if a plane doesn't arrive on its scheduled time, the airline company cannot blame Einstein. Jet planes at top speed move a million times slower than the speed of light, so that the time dilation is immeasurably small for the average passenger.

Third: what we call 'simultaneous' must be completely revised in the relativistic world. Time is not the 'steadily flowing river' of classical mechanics. The speed of light is the maximum speed in our Universe. If the speed of the clock with respect to an observer equals the speed of light, that observer would see the light ray reach the next mirror only after an infinite amount of time. In other words, the clock stands still. If the clock's speed could be greater than the speed of light, the light ray would never catch up with the mirrors of the clock, and the clock would stand stiller than still — which is absurd.

This somewhat woolly argument can be made more precise by noting that, if the speed of the clock is bigger than c, the mathematical expression for the time dilation would contain the square root of a negative number. That is permitted in math, but the resulting numbers cannot be ordered, and as far as we know, it is possible to order time by determining what happens first and what happens next.

← Light paths in a Lorentz clock. The orange arrow shows the up-and-down path of the light when the clock is standing still with respect to the observer. When the clock is moving (white arrow) the light path is a sawtooth-shaped line (yellow arrows). Because the yellow diagonal is longer than the orange path, and the speed of light is always the same, the crossing time of the light is seen to be bigger in the moving clock than in the same clock that is standing still with respect to the observer. That is where relativity got its name: because the speed of light is absolute, time is relative.

Gravity Does Not Exist

The speed of light is always the same. Because this speed is absolute, time is relative: a moving clock is seen to tick more slowly than the same clock standing still with respect to the observer. This is called time dilation, and the equation describing it shows that the speed of light is the maximum speed attainable in the Universe.

Consequently, instantaneous actions or connections over a finite distance are impossible. In our Universe, all things are always under way, whatever they are. No two events in space-time may be linked instantly; the news that something has happened always takes some time before it has reached other places.

Likewise, the properties of space-time itself cannot be linked instantly across finite distances. This implies that space-time properties may vary from place to place and from time to time. If, for some reason, space-time were not exactly uniform, meaning that its properties differed from event to event, it would be impossible to smooth everything out instantaneously.

When the structure of space can vary in the course of time, we are justified in saying that space-time has its own dynamical behaviour, so that space-time may be seen as real stuff with its own structure, like the Oude Kerk in Delft. Unlike that old church, space-time doesn't just stand

→ Illustration of two-dimensional curved space. The background is made black to symbolize that it is not part of the two-dimensional world. The hills and valleys of the surface point in an abstract third dimension, a direction to which there is no access because the two-dimensional surface is all there is. The image does not have a specific orientation on the page. That may seem odd, but there is no 'up' in the space of our Universe. Any specific direction is equivalent to any other.

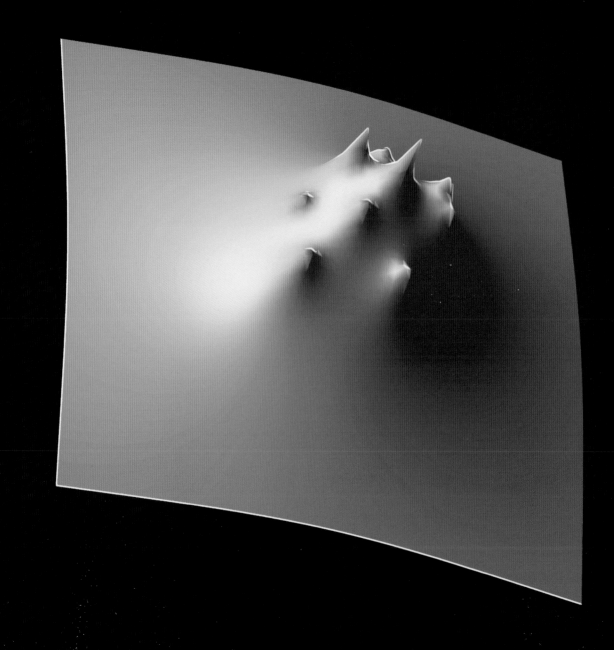

there, but it is dynamic. Space-time is not some sort of invisible graph paper on which the paths of all things are drawn, the way Newton said it is. Space-time is stuff, with a dynamic structure.

The invention of the mathematical equation that expresses this astonishing result was Einstein's greatest contribution to physics, even greater than his other enormous achievements, according to just about all physicists. The formulae have a stark beauty of their own,[55] but they need not be presented here. Translated into plain language, they state that *the structure and dynamics of space-time are determined by the arrangement of mass, energy and momentum.*[56]

Astonishing indeed, because it follows from this equation that, first, gravity does not exist; second, it is now clear why Stevin's experiment showed what it did.

If there is no such thing as gravity, what then is the reason why the orbits of the planets are curved? What causes the acceleration that this curvature implies, according to Huygens's formula for the centrifugal effect? The answer is technically extremely complicated, but simple to summarize: the apparent acceleration is not caused by a force. Curved space-time produces curved orbits in the form of the shortest paths through its (3+1)-dimensional landscape.[57]

← Curved space produces curved paths, also for light rays. The bending of light by the structure of space-time is called *gravitational lens effect*. The smears and streaks in this image of the deep Universe are actually images of very distant galaxies. The images are distorted by the space-time curvature in the cluster of galaxies between us and those galaxies far behind the cluster.

55 Sander Bais, *The Equations, Icons of Knowledge*, Harvard University Press 2005, p. 65.
56 In classical mechanics, the momentum of an object is its velocity multiplied by its mass.
57 Actually, the requirement for finding paths is that they are *extremal*, i.e. either the shortest or the longest according to a particular distance recipe. But we will not need that subtlety here.

Most popular explanations of space-time orbits have balls rolling down hillscapes, and that is one hundred per cent wrong. In an analogy like this, there is no such thing as 'rolling down the surface', for the excellent reason that this surface is all there is in the dynamics of General Relativity. The curvature of the paths is not due to an external agent 'pulling' them over hill and valley; they are simply the shortest paths in a curved space-time.

Space-time is real stuff, every bit as real as bricks and mortar, so we can build space-time structures. In the full (3+1)-dimensional world this is difficult to visualize, so let us simplify everything to a two-dimensional world. I earnestly ask the reader to do the experiment

→ Construction of a curved two-dimensional space by cutting and pasting. Start with a sheet of paper. In this flat world, the shortest connections between points are straight lines. Draw two parallel lines, about 10 centimetres apart. These represent paths in the two-dimensional world. Cut the paper halfway between the lines, move one edge of the cut over the other, and paste it to form a cone. Effectively, that means removing a wedge-shaped piece of space from this two-dimensional world. In the resulting curved space, parallel lines cross, and curved space is seen to give curved orbits. The drawn lines are still the shortest connection between points. The curvature can be measured by noting that the circumference of a circle, divided by its diameter, is less than π if the circle encloses the tip of the removed wedge. Einstein's great discovery was that matter in (3+1)-dimensional space-time produces a curvature. In this two-dimensional example, the matter that makes the surface conical instead of flat is concentrated at the end point of the cut in the paper (red dot).

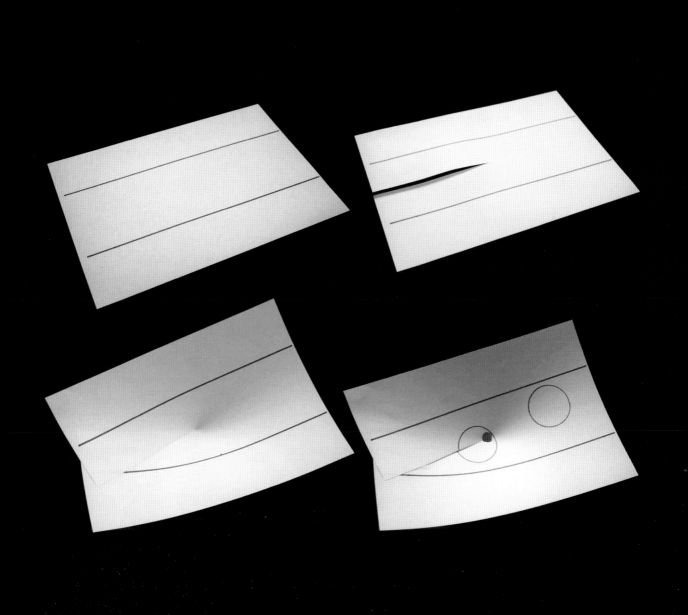

shown in the illustration. It takes no more than two minutes, whereas it may well take two years to master the mathematics necessary to describe the geometrical structure of four-dimensional space-time.

By changing the structure of the paper on which straight lines are drawn, the lines become curved. They are still parallel, and may even cross, though the lines themselves remain untouched. Only the shape of the two-dimensional space in which they lie has been changed. Parallel lines do not cross in a flat space. But the cut-and-pasted space is curved, and *curved space gives curved orbits*. Their curvature is not due to Newtonian gravity, nor indeed to any other force: the shape of the paths is a consequence of the shape of the space in which they lie.

All this was pointed out by Einstein in 1916. His most dramatic insight, however, did not concern the description of the curvature of space-time. Techniques for handling curved spaces had been under development by mathematicians, starting with the Arabic geometer al-Jayyani who explored the properties of triangles on a sphere.[58] Einstein discovered something in physics, namely: *matter curves space-time*. The end point of the cut in the paper described above is special: this is where the matter is. The stuff of the Sun curves its surrounding space-time. The motion of objects in that space is fixed by the requirement that they must follow the shortest paths. The length of a path is not determined by the properties of the moving objects, but is purely due to the structure of the space in which everything moves.

Only then, 331 years after Burgomaster De Groot climbed that tower in Delft, was Stevin's experiment understood: *gravity does not exist*. Newton could only make his mechanics work by requiring that the ac-

58 Abu abd Allah Mohammed ibn Muadh al-Jayyani (989-1079), mathematician from al-Andalus (Spain), author of the first known publication on spherical geometry.

→ In places where the concentration of mass, energy and momentum is large, space-time is strongly curved. The core of a star cluster is such a place.

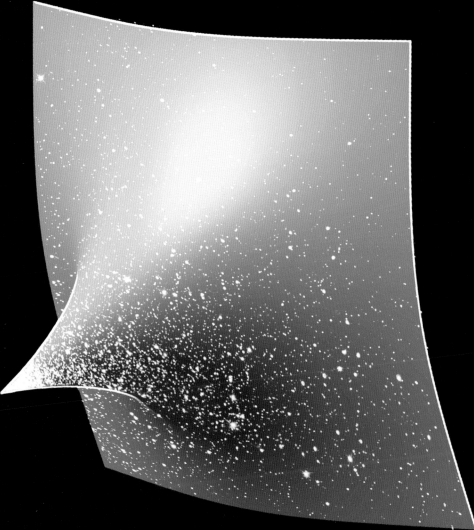

celeration caused by his 'universal force' be independent of the mass of the accelerated object.[59] In General Relativity, orbits are curved because of the structure of space-time. The orbit of an object moving in space-time is a consequence of that structure, which is the same for all objects.

An experiment done in 1585 cannot be the final word on any subject. Many new tests with increasing subtlety and precision were invented to see if there are properties of matter that influence their acceleration due to gravity.[6] In the most sensitive of these, two balls of different mass and composition were fixed to the ends of a rod, forming a sort of 'dumbbell'. This dumbbell was then made to oscillate on a quartz fibre, so that the balls alternated their position with respect to Earth and the other objects in our solar system. Should the acceleration of gravity on one ball be different from that on the other, this would have observable consequences for the way in which the dumbbell oscillates. The remarkable fact is that, to this very day, no dependence of the gravitational acceleration on material properties has been found, even after precise measuring to 13 decimal places.

If Stevin's experiment had shown Aristotle to be right, namely that heavier objects fall faster than light ones, then it would have been an enigma that the Moon and Earth remain so close together as they orbit the Sun. The mass of the Moon is 81.3 times smaller than that of Earth. If the interaction with the Sun did depend on mass, our satellite would soon have disappeared into space. But since Earth and its Moon are so close together, the space curvature due to the mass of the Sun is nearly the same for both of them, so that they have remained bound together for 4.6 billion years.

Today, we often see Stevin's experiment in a different form, in the

59 The force of Newtonian gravity between two interacting objects is proportional to the product of their masses. The fact that the force depends on both masses indicated to Newton's contemporaries that the objects must somehow 'know' about each other's properties. That looks rather bizarre, so Newton's proposal was met with skepticism. Because a force is the product of mass and acceleration, the mass of the accelerated object does not occur in its equation of motion.

rather clumsy spectacle of an astronaut floating next to a spacecraft. In the media this is called 'weightlessness', a confusing term. What actually happens is that both follow almost exactly the same trajectory, which is determined by the geometry of space-time, not by some sort of force.[60]

The orbits of Earth and all other objects in our solar system are curved because the Sun curves the space-time surrounding it. That curvature is imposed by our star, according to the Einstein equation: that says the structure and dynamics of space-time are determined by the arrangement of mass, energy and momentum.

Gravity does not exist. 'Gravity' is a historical term for the consequences of the structure of space-time. But we are stuck with the word, which may be just as well. After all, the 'sunset' doesn't exist either, as we have known ever since the discovery of the rotation of the Earth. Nevertheless, we can still enjoy the view just as much as we ever did, knowing what we know now.

60 The difference between adjacent orbits is very small, but measurable. For example, the difference in the space-time curvature that the Moon causes across the body of Earth produces the tides in Earth's oceans. Likewise, tides caused by Earth across the Moon lock the rotation of the Moon in such a way that it always points the same side towards Earth.

Reflections

The main reason why mechanics was a very good candidate for the title 'Theory of Everything' is probably that it feels so definitive, not only in its strict and transparent mathematical structure, but also in its workings. If we want to know the future, all we have to do is specify the positions and velocities of all material at some initial time, and then the remainder of eternity can be computed, at least in theory. With a theory of everything, we can say with Leibniz: let's calculate! *Calculemus!* Voltaire wrote:

> *All occurrences are produced by one another. [...] Under the same circumstances, the same causes produce the same effects.*[61]

Einstein's relativity theories did not change that; again, at least in principle. The computations become vastly more complicated, but the fact remains that, in relativistic mechanics, the future follows uniquely from the past.

Everyday mechanical events usually show that this is an acceptable point of view. In practice, the fact that the Universe consists of an unruly number of particles makes that initial specification and its subsequent computation difficult, but not necessarily impossible (plus the problem that the computer must be part of what it computes, but never mind that).

However, there is a danger lurking under this seemingly calm surface. Because mechanics demands that we specify the whereabouts of all material in the Universe at an initial time, the question arises what that 'material' is and, in particular, what its internal constitution is.

61 Voltaire, pseudonym of François-Marie Arouet (1694-1778), French writer and philosopher: *Dictionnaire philosophique, ou la Raison par alphabet*, 1764.

It had long been clear that gravity cannot be the only force in the Universe, because gravity is a purely attracting force, so it cannot produce stable objects by itself. But research into electricity had shown that an electric charge can be positive as well as negative, and that charges with the same sign repel each other. The experiments and theories of Faraday and Maxwell, developed in the 19th century, showed that the forces of electricity and magnetism can be in equilibrium.[62] Therefore, electromagnetic forces were good candidates for the explanation of the structure of matter.

Then, in the first quarter of the 20th century, it was discovered that matter is not continuous, but built out of discrete particles[12]. These particles form families, analogous to the well-established 'chemical elements', and within a given family the particles are precisely identical. Faraday's electric current turned out to be a stream of an immense number of electrons. Rutherford discovered that the atoms of chemical elements are built from a nucleus that has a number of electrons bound to it, in a swarm that is some hundred thousand times bigger than that nucleus.[63]

It was known that electrons cannot orbit in atoms like Newtonian mini-planets, because Maxwell's theory predicted that they would lose their energy due to the fierce emission of radiation, and crash into the nucleus in no time at all.

Thus, while the falling of Stevin's leaden balls had been beautifully explained by Einstein's space-time curvature, the structure of the lead and all other matter remained an unsolved problem. We will presently see that a solution for this was soon found, but that the description of the behaviour of mass is totally at variance with what is known about space-time.

The key experiment, in this case, does not require a city official to

62 Michael Faraday (1791-1867), English physicist; James Clerk Maxwell (1831-1879), Scottish physicist.
63 Ernest Rutherford (1871-1937), New Zealander physicist.

climb a tower. In fact, the effect is so familiar that everyone has seen it.

When you stand in a brightly lit room in front of a window in the evening, you can see your own reflection in the glass. At the same time, someone standing outside can see you, too. To a perceptive person, this is very odd: the light that is reflected back towards your eyes leaves your face under precisely the same conditions as does the light that passes through the glass to the person outside.

Of course Huygens knew this, too, and his shock-wave theory neatly solved the problem. The light does not choose anything: part of the wave passes through the glass, and another part is bounced back. Huygens must have observed this many times in his native Holland, where waves on the water can be seen to partially reflect from obstacles hidden under the surface. Maxwell's theory seemed to clinch the point.

But then it was discovered that light is made of particles, just like matter. A beam of light is in fact a stream of a vast number of light particles, *photons*, just as an electric current is a stream of electrons.

It is as if the light makes a 'choice' about whether it can go through the glass or be reflected. What happens when a photon interacts with the windowpane? Does it bounce back, or does it go through? The photon that is reflected towards your eyes leaves your face under precisely the same conditions as does the photon that passes through the

→ Hypothetical experiment in which the position of an electron in a hydrogen atom is determined 10,000 times. Such a large number of measurements produces a 'point cloud' that shows where the electron is most likely to be found. The nucleus of the hydrogen atom (a proton) is in the centre of the image. The distribution shown is the (322) state of the H-atom. The colour shows the distance: blue is far from the atomic nucleus, yellow is near.

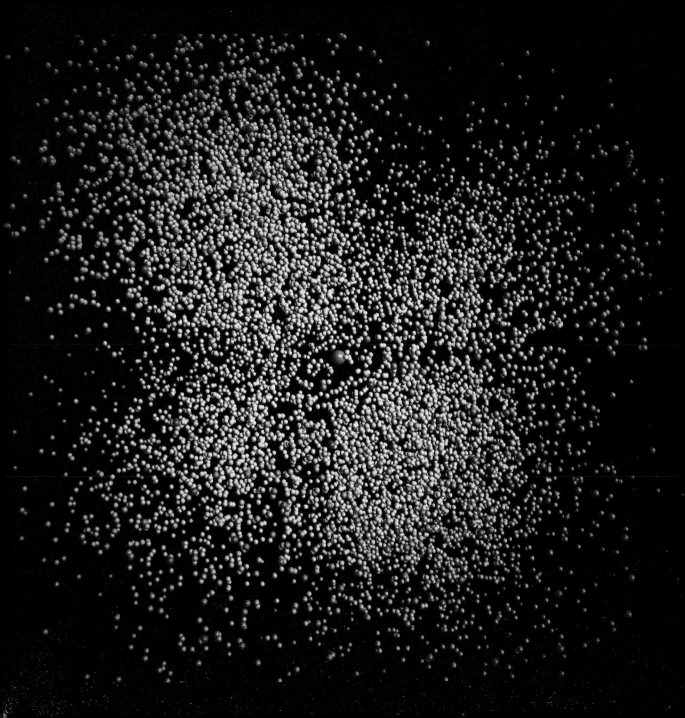

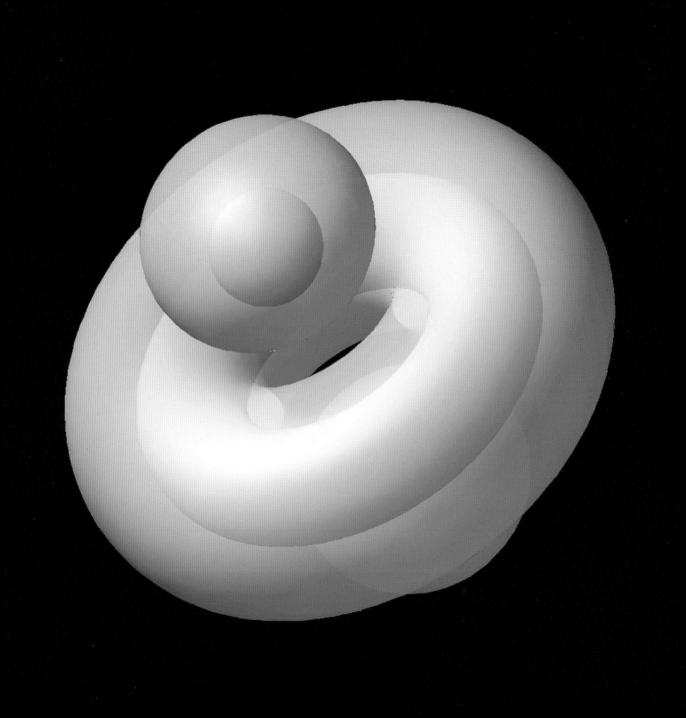

glass to the person outside. The conclusion must be that Voltaire and his common sense are wrong. In our Universe, *the same causes do not always have the same effects*.

Consequently, the behaviour of 'elementary' particles is extremely different from that of large objects such as marbles.[12] That is why I will often use the term *quantum particles*. Depicting a hydrogen atom as if it were a miniature billiard ball (or, even worse, a solar system) is grossly wrong and leaves an entirely false impression. It is possible to approximate what an atom looks like, but it takes a while to get accustomed to it.

Perhaps the closest one can come to the physical truth is by showing the result of a hypothetical experiment that determines the position of an electron in an atom. A single such measurement could find the electron anywhere, maybe even near the star Arcturus. But a large number of measurements produces a 'point cloud' that gives a fair impression of where the electron is most likely to be found. Not in the centre, not moving, not at infinity, but on average somewhere in between, very close to the nucleus. In no way does this resemble planetary orbits.[64] An alternative is drawing surfaces in space that correspond to the probability of finding an electron there. This is analogous to altitude lines drawn on a topographic map.

← Two surfaces of the probability distribution that is shown above as a point cloud. The nucleus of the hydrogen atom (a proton) is in the centre of the image. The electron has a 90% probability of being found in the volume of space enclosed by the outer surfaces (two balls and a ring). There is a 50% probability of finding the electron in the volume enclosed by the inner surfaces (smaller balls and ring).

64 I find it totally bizarre that the logos of many scientific organizations in the field of atomic and particle physics contain orbital ellipses.

It was soon established that not only electrons, but all particles behave in this way. This very odd behaviour required the construction of a totally new dynamical theory. The collective name for particles (electrons, photons and the rest) is *quantum*, expressing the discrete nature of the building blocks of matter. Accordingly, the theory that was devised to describe this behaviour was called *quantum mechanics*.

The mathematical formulation of quantum mechanics that most clearly illustrates the enigmatic 'choice' behaviour at the windowpane is due to Dirac.[65] It shows emphatically how different quantum mechanics is from the classical theory of motion. Because the same causes are not always followed by the same effects, it is no longer sufficient to specify initial conditions and then let the future do its thing. Instead, both the initial and the final conditions of an experiment or interaction must be specified.

In quantum mechanics we do not compute the future from the past, but instead we compute the *probability* that a given past is connected to a given future. Past and future must *both* be specified. In graphical form, Dirac's formula reads:

probability = <past||interaction||future>

so that a given initial state of the Universe, summarized by the expression <past|, is connected to a specific final state |future>. This connection is due to a coupling mechanism, summarized in print by the form |interaction|. For example, the windowpane experiment requires the calculation of two probabilities, namely

reflection = <photon in||glass||photon bounced back>
transmission = <photon in||glass||photon passed through>

65 Paul Adrien Maurice Dirac (1902-1984), English physicist.

From expressions such as these, all probabilities that are relevant to a given setup can be derived. All we know is, that the total probability adds up to 1 (or 100%, whatever one prefers).

In general, there are infinitely many possibilities for the connection between the <past| and the |future>. In quantum mechanics, we say that the past (usually called the <in| state) is connected to the future (the |out> state) by a whole range of *alternatives*.

Jes' Rollin' Along

Computing the net result of all possible alternatives is a monumental task, even in apparently simple cases. This is not a physics textbook, but it may be useful here to give some indication of how quantum motion is computed. This will show how far from everyday experience quantum behaviour is, and how unlike the classical motion we saw in Galileo's parabola and Huygens's spinning circle.

Motus inter corpora relativus tantum est, said Huygens: movement between objects is relative in all aspects. The absolute velocity of an object does not exist. Any dispute about the 'true' state of motion of a single object is meaningless. In particular, it makes no sense to distinguish between 'rest' and 'motion', provided that this motion proceeds with constant velocity.

A classical particle moves with constant velocity in the absence of accelerations, which are caused by forces. But what does a quantum do?

→ Addition of two quantum paths. Place the quantum wheel at the beginning of one path (first strip). Roll it all the way to the end (second strip). Note the phase of the wheel (position of the yellow arrow). Then place the wheel at the beginning of the other path (third strip). Roll it all the way to the end (fourth strip). Note the phase of the wheel (position of the blue arrow). Add the two arrows together by linking them up, tip to tail (bottom strip). The resulting amplitude (white arrow) is a measure of the probability of the whole process: the probability is equal to the square of the length of the arrow; its direction doesn't count. In this example, the process is made up of just two paths. In realistic cases, there are infinitely many of them, so that actual computations of this type are very complicated.

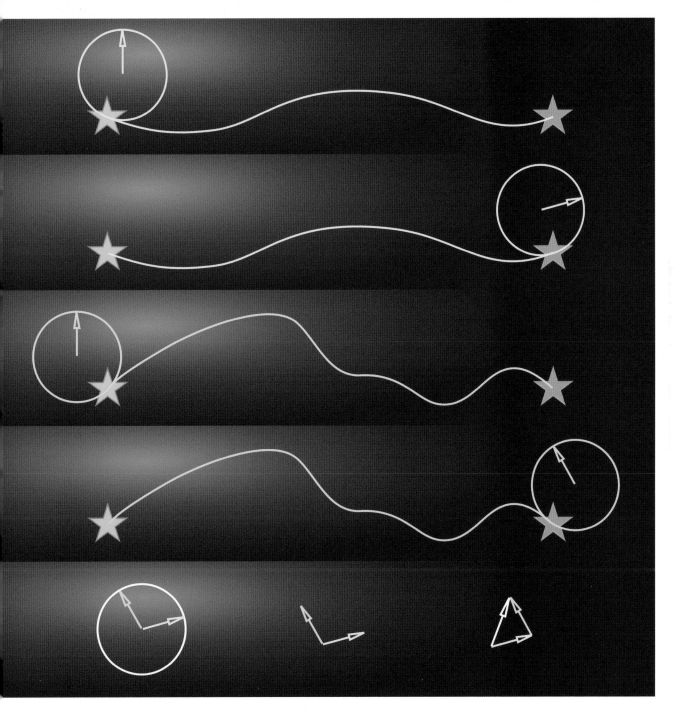

Picking just one path for the quantum to follow makes no sense in our Universe, because the windowpane experiment tells us that the same causes do not always have the same effects. In fact, we must do exactly the opposite: each path that is not explicitly forbidden, and that is not distinguished from other paths in any way, must be allowed. All force-free paths are to be considered collectively as equal-rights alternatives.

Now we are faced with the following problem. If all paths have equal rights, do they all have equal probability? It turns out that the answer is 'no'. How must we compute the various probabilities? The answer to that is rather curious and totally contrary to everyday intuition. This is so bizarre, that it took scientists something like half a century to work it out.

A quantum on a path behaves like a rolling wheel. Imagine that we have a wheel with a given circumference.[66] On the wheel, an arrow drawn from the axle to the air valve on the rim is called the *amplitude* of the quantum. When the wheel rolls along, that arrow rotates about the axle. The angle through which the wheel has rotated is called the *phase* of the quantum.

The task now before us is the following. First, fix the beginning and the end of a quantum path (corresponding to the <in| and |out> states above). Second, construct all paths from <in| to |out>. Third, roll the quantum 'wheel' from the beginning to the end, starting with the arrow always in the same position (straight up in the drawing, arbitrarily called 'phase zero'). Fourth, mark the direction of the arrow on the wheel at the endpoint; this arrow is the amplitude associated with the path.[67] Fifth, do this for all paths. Sixth, find the amplitude of every path. Seventh, calculate the total amplitude over all paths (alternatives)

66 This length around the wheel is sometimes called the 'wavelength' of the quantum. In the case of light quanta (photons), this length corresponds to the colour: blue light has a short 'wave' length, red is long.

67 Thus, any whole number of rotations that the wheel has turned along the way makes no difference at the end. For example, if the wheel has turned 15 and 1/6 times, the phase at the end is 1/6 of a turn, that is an angle of 30 degrees.

by adding the arrows: just link them all, tip to tail, in any order what-ever.

The length of the resulting arrow (not its direction) is a measure of the probability of the whole process, with every alternative duly in-cluded. The existence of alternatives was expressed above in the form of 'the same causes do not always have the same effects'. In the picture shown above, there is an infinity of alternatives between the depar-ture of the quantum and its arrival. This makes quantum mechanics (sometimes called *particle mechanics*) radically different from classical mechanics. Once an astronomer has determined the orbit of a planet, that work is finished. But the 'orbit' of a single elementary particle has no meaning at all.

The phase of a particle is not directly observable; only phase differ-ences count. Having read Huygens's relativity principle, that should not sound so strange. The position and the velocity of an object are not observable; they can be determined only with respect to other objects, because *motus inter corpora relativus tantum est*. Had Huygens known about quanta, he might have written something like *phasus viarum rela-tivus tantum est*: 'the phase along paths is relative in all aspects'.

The wheel-rolling 'phase behaviour' causes quanta to act like waves, to a certain extent. At the end of the path, only phase *differences* count: every whole turn of the wheel is discounted. Thus, there is nothing that distinguishes one revolution of the phase wheel from another one. The phase behaviour is strictly periodic. In our large-scale world, that periodicity shows up in the form of wave-like behaviour. For example, the length of the circumference of the phase wheel is commonly called the 'wavelength'. That term is still widely used, even though a quantum is not a wave. But never mind — gravity doesn't exist either.

That is why Huygens's theory of light was so successful, even though we now know that light is a stream of photons and not a propagating shock. It is sometimes said that a quantum is 'alternately a wave or a particle', but that is wrong. *A quantum is a particle with phase.* If that sounds strange at first, think about *a neutron is a particle with mass*, or else *an electron is a particle with charge*. These sentences are not difficult, but they are strange — of course, because it's physics.

In our large-scale world, the periodicity associated with quanta shows up when some of the paths of the quanta are blocked, in a way that is similar to the picture above. If the propagation of the quanta is tampered with at random, nothing much changes, except for an overall decrease and smearing-out of probability. But if the quantum paths must pass over a regular obstruction, such as the very fine grooves made by a swipe with sandpaper over a polished metal surface, the effect of the phase becomes apparent. The metal then shimmers with various colours, because the photons that bounce off the parallel scratches have their phases lined up. The effect depends on the wavelength. Our eye

→ Probability distribution when the quantum (coming in from the right) is constrained to move through two openings in a wall. The height of the 'landscape' indicates the probability that the quantum is found in that place: blue is low, gray is high. The presence of the wall openings produces several zones where the probability is very small, radiating outward in a fan-like shape from a place between the two openings. In between, there are wave-like zones where the quantum is more likely to be found. The image looks complicated, and it is, but it was computed with exactly the same wheel-rolling procedure as was shown in the previous figure. Patterns like these are seen in many common objects such as soap bubbles, thin films of liquid or plastic, and hologram images on passports or bank cards.

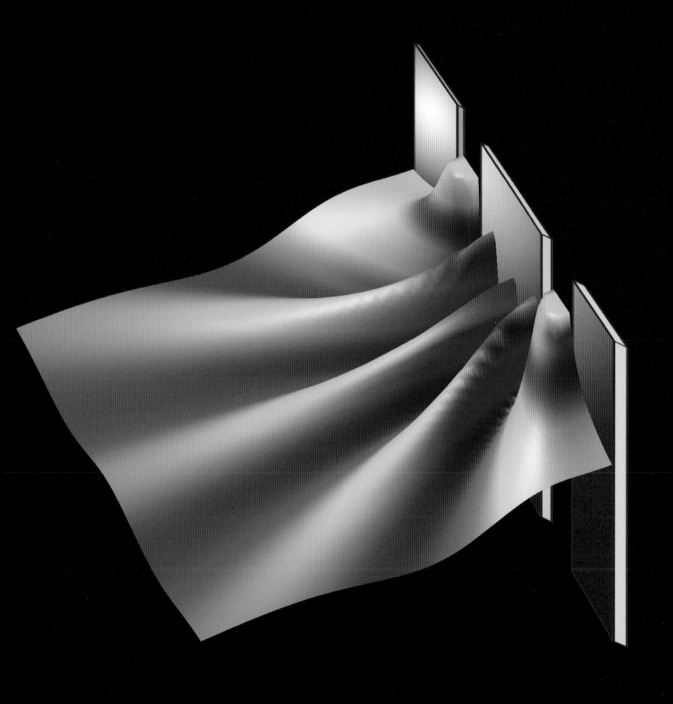

interprets wavelength as colour: blue corresponds to short and red to long wavelengths. All photons together produce iridescent colours, for example, in soap bubbles.

Historically, the discovery of the phase behaviour of quanta took about three centuries. Huygens took the first step in his work on the propagation of light.[68] With his usual perceptiveness, he noted several properties of light which told him that a beam of light cannot be like a stream of water or air, for example, the fact that beams of light can cross without colliding. He explained light as a kind of shock or percussion that propagates in an unseen 'aether'.

Two centuries later, Maxwell described light as the wave-like vibration of an electromagnetic field. When Planck tried to calculate what a box full of such vibrating fields does, he did not get anywhere until he assumed that the energy of the vibrations comes in chunks, which were called *quanta*.[69]

The 'quantization' of the electromagnetic field produced excellent theoretical results, but the question remained: are quanta real things, or just a computing trick? Einstein and Bohr, with the daring and perceptiveness of brilliant youth, took the attitude: well, let's assume that quanta are real, and explore the consequences. That led to another surge of splendid results.[70]

However, the Planck-Einstein-Bohr quantum theory said nothing about the way quanta move. Schrödinger invented a description of quanta in motion.[71] His *quantum mechanics* was a monumental advance in the understanding of the mechanisms of the Universe. The mathematical formulation that Schrödinger invented hinged on the 'phase' property of particles. The mathematical waveforms that scientists had used to describe all sorts of oscillations and vibrations, such as waves in air or on water, turned out to be just as useful in the description of quanta.

68 Christiaan Huygens: *Traité de la lumière*, Pieter van der Aa, Leiden 1690.

69 Max Karl Ernst Ludwig Planck (1858-1947), German physicist.

70 Niels Henrik David Bohr (1885-1962), Danish physicist.

71 Erwin Rudolf Josef Alexander Schrödinger (1887-1961), Austrian physicist.

Unfortunately, this led many people to use the term 'wave mechanics' instead of 'quantum mechanics', even though a quantum is not a wave. The fact that the rotating arrow on the phase wheel traces a wave-like shape is a mathematical coincidence.

Meanwhile, Einstein had shown that the Universe is relativistic in a way that differs from Huygens's relativity. Einsteinian relativity was built on the finding that the speed of light is always the same, independent of the motion of the emitter or the receiver. Dirac brought the quantum picture up to date by constructing a quantum theory that is relativistic. The importance of being relativistic will be explored below in *A Twist to the Tale*. First, however, we have to consider something that every dynamical theory must deal with: the interaction between particles.

Feynman's Web

Stevin's leaden ball experiment, the theme that runs through this story, has now broken up into two parts: the equal acceleration of the two different masses, and the structure and behaviour of matter. The equal-acceleration behaviour of 'gravity' is explained by the properties of space-time discovered by Einstein. Next on the agenda are those leaden balls or, more generally, any matter.

To understand the structure and behaviour of matter, we must delve a bit more deeply into the world of particle physics that is too small to be seen with an optical microscope. This will give us more insight into the nature of forces, and the possible beginning of a connection between the space-time world at large with the particle world on small scales.

The interaction of light with matter, such as in the case of the glass of a windowpane, takes place via an infinity of alternatives. These alternatives, when added together by means of the rolling-wheel process, produce effects collectively known as *interference*. For example, the 'choices' that light rays have in reflection-and-transmission at surfaces (such as glass or water) can produce visible interference, as they do when producing the shimmering colours in a soap bubble.

With a little effort it can be shown that, in empty space, the most probable path of a quantum is a straight line. At least this aspect of Huygens's relativity is nicely built into quantum mechanics. But what about curved paths? How are accelerations introduced?

Let us go back to the presentation of the rolling phase wheel associated with the motion of a quantum. The shape of the most probable path is determined by the addition of the phases of all possible paths between two points. It stands to reason, then, to introduce accelerations (and thus curved paths) by means of a prescription of the phases along the paths.

If the shape of the most probable path is curved, Huygens would have said that the object moving along that path has been accelerated. In the quantum world, the shape of the most probable path is governed by the arrow-addition of the phases of the quantum wheel along all possible paths. Thus, what we experience as an acceleration in our large-scale world is due to the behaviour of the phases of the quanta.

Feynman[72] created a method by which one may exhaustively list all possible ways in which such a phase change can be made. He did this in a way that looks amazingly intuitive. In the Feynman procedure, an encounter between particles is described by a point in space-time at which the phase wheel gets an extra twist. Feynman called such an encounter point a *vertex*.

The most probable path of a quantum depends on the phases of all possible paths through space-time. Thus, the most probable path may be curved if the phases are suitably adjusted. Curvature of a path means that the particle was accelerated (remember Huygens's centrifugal effect). Therefore, a prescription of these phases amounts to a prescription of the acceleration.

It turns out that the most primitive quantum interaction is a direct connection between three particles belonging to two families: the *fermions* and the *bosons*. Fermions were named after the physicist Fermi,[73] bosons after his colleague Bose.[74]

Two identical fermions can never share the same dynamical properties, such as their position and their velocity in space-time. If two such fermions are confined to a small volume of space at a given time, they must have different velocities. Therefore it is impossible to keep them together in a very small region, as if they were quarrelsome children. A collection

72 Richard Phillips Feynman (1918-1988), American physicist.
73 Enrico Fermi (1901-1954), Italian physicist.
74 Satyendra Nath Bose (1894-1974), Indian physicist.

of fermions always forms an object with a finite size, because a fermion does not 'tolerate' others nearby. Fermions of a given type must keep a certain minimum distance from each other, so that they become something big when there are very many of them: gas clouds, bricks, oceans, stars. That is why these particles are known as *matter* in daily life.

Identical bosons behave in exactly opposite ways: bosons have a tendency to assume the same dynamical state, flocking together and forming a coherent whole. This makes bosons carriers of what we call *forces* in our everyday world.

The interaction between these particles arises when two fermions are connected to one boson[75] at a point in space-time; this event is called a *vertex*. A boson is comparable to a pizza deliverer, with a box on the back of a motorbike containing whatever is supposed to be delivered upon arrival. Thus, the contents of that box determine the nature of the interaction.

When all interactions are summed up into a grand total that is visible in our world, much larger than the sub-micro-world of the quanta, we speak of a 'force'. The properties of a force are determined by the way in which its boson is coupled to the fermion at the vertex.[76] For example, a magnet is a chunk of fermions (mostly atoms of iron), a magnetic field is a stream of bosons (left- or right-spinning photons).

In the particle world, a fermion interacts with another fermion through the exchange of a boson. This is comparable to a tennis player (fermion) who hits a ball (boson) towards another player (fermion). If this

← A single Feynman vertex, represented by the fuzzy blue-white ball. At the vertex, three particles are connected. The phases of the quanta change at the vertex.

75 What types of particles can connect at a vertex is determined by the 'spin' of the particles. The discussion of spin (the amount of internal rotation of a quantum) is beyond the scope of this book.

76 For more details, see my book *The Force of Symmetry*, Cambridge University Press, 1999.

happens in free space, the first player will bounce back upon hitting the ball. When the second player receives the ball his/her velocity will change also. In this way, each player feels an acceleration due to the exchange of the ball.

This process is summarized graphically by a *Feynman diagram*, the most basic of which is a drawing in the form of the letter H. Each vertical branch symbolizes a fermion, the horizontal bar connecting them represents a boson.[77] There are two vertices, where each fermion connects with the boson that is exchanged between the fermions. Feynman assigned a specific mathematical expression to each part of such a diagram: one for the incoming fermion on the left, one for that fermion going out, likewise for the fermion on the right, one for each vertex, and one for the boson. The resulting formula is used to compute the probability that the given past (incoming fermions) is connected to the specified future (outgoing fermions).

The elementary particles called electrons are fermions; photons are bosons. The interactions between them produce the phenomena of electromagnetism in our everyday world. Quarks are fermions, gluons are bosons. The exchange of gluons between quarks produces a force that binds the quarks into protons and neutrons, the particles that make up atomic nuclei.

→ The bronze and magenta branches in this pictorial Feynman diagram symbolize two fermions, the blue tube connecting them represents a boson. Each fermion connects at a vertex (fuzzy ball) with the boson that is exchanged between them. Each part of a Feynman diagram is an alternative similar to the wheel-paths shown above. The whole diagram is one alternative of a specific interaction.

77 Notice how very different this is from the Newtonian concept. Newton's 'universal force' between two objects depends on both their masses. This suggests that the objects must somehow 'know' about each other's properties. That looks rather bizarre. But the Feynman diagram shows that the interaction between quanta is due to an actual connection between the particles. The appearance of a boson exchanged between vertices implies that a quantum interaction is not an 'action at a distance'.

A stream of free photons we call 'light'; the exchange of photons between electrons produces the electromagnetic force that binds electrons to atomic nuclei, and binds atoms in chemical compounds. Feynman described more details of this theory, *quantum electrodynamics*, in his brilliant book, *QED*.[78] This type of description, called *quantum field theory* (QFT), is used for all known particle interactions.

The H-shaped Feynman diagram is the simplest, but by constructing more and more intermediary vertices and boson lines, we may build up a multiply infinite network of interactions.[79] Taming this unruly jungle has been very difficult but ultimately successful, thanks to the splendid work of physicists such as Yang, Mills, Veltman and 't Hooft.[80] The whole material world is a network of these interactions. 'Matter' and 'force' are historical names for the large-scale effects of that immense web of smaller-than-microscopic quanta.

78 Richard Phillips Feynman, *QED, the Strange Theory of Light and Matter*, Princeton University Press, 1985.

79 In textbooks, the various branches are drawn as lines. In my illustrations I use a fancy shape to remind the reader that the lines in a Feynman diagram do not represent particle paths, and that the particles may not be 'elementary'.

80 Chen Ning Yang (1922-), Robert Mills (1927-1999), Martinus Veltman (1931-) and Gerard 't Hooft (1946-); Chinese, American and Dutch physicists, respectively.

→ Objects that in chemistry or atomic physics are treated as single particles, do in fact have a very intricate structure. For example, a helium nucleus is built out of two neutrons and two protons. These, in turn, are made of quarks, bound together by the exchange of gluons. To emphasize how far this is removed from 'balls', I use baroque shapes for quantum particles in illustrations.

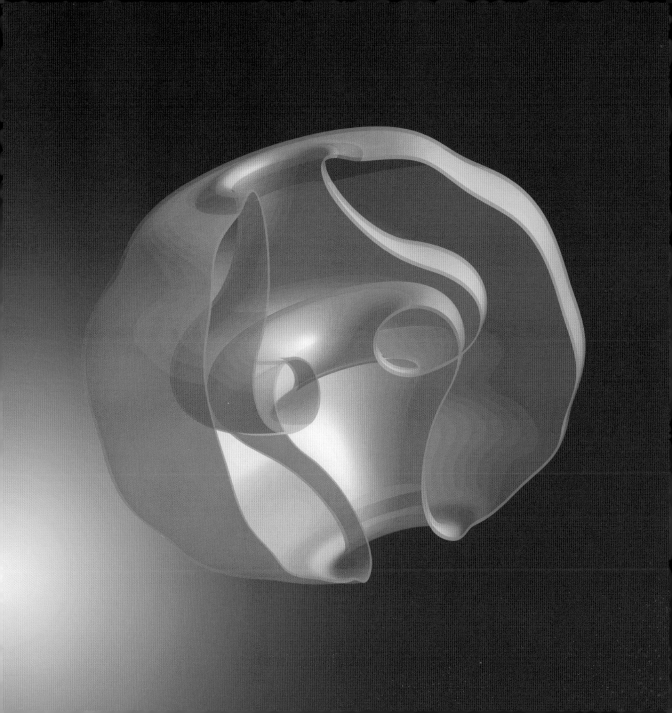

A Twist to the Tale

The paper-cutting experiment above illustrated Einstein's General Theory of Relativity. It showed that the motion of an object is determined by the structure of space-time: curved space gives curved orbits. The relationship between matter and space-time is mutual. On the one hand, the orbit of an object is due to the structure of space-time. On the other hand, that structure is determined by the arrangement of the mass-energy-momentum of matter. We also saw that the state of motion of quantum particles is determined by the coupling at a Feynman vertex.

This raises the question: *what determines this coupling?* Or, in the language of our large-scale world: what determines the properties of forces?

The observation of reflection-and-transmission of light in a window-pane shows that quantum particles behave in ways that are radically different from the motions of big objects. Orbits in classical mechanics are fixed when all initial positions and velocities of the particles are prescribed together with a recipe for the accelerations. In quantum mechanics, we must determine the transition probability that connects a given initial state to a specified final state. These probabilities are computed by means of the Feynman diagrams introduced above.

The key to interaction in a Feynman diagram is the vertex, the point in space-time (called an *event* in relativity theory) at which particles are coupled. The interactions at a vertex are determined by *symmetries*. Before discussing how this works, let us look a little more closely at what is meant by 'symmetry'.

A symmetry may be loosely described as 'a change that leaves something unchanged'. Rotation is an example of a symmetry. A sphere is unchanged when it is rotated about its centre. Mathematicians say that a sphere is *invariant under rotations*. You might say that a sphere is in fact *produced* by a symmetry, because a sphere is the set of all points that have the same distance to a given point, and that 'same distance' is the 'something that does not change'.

Galileo thought that the rotational symmetry of a circle implies that the 'natural motion' of planetary orbits is composed of circles, but Huygens proved that circular motion requires an acceleration. As we saw above, Huygens formulated a very different symmetry: the universe remains unchanged if everything in it were displaced over a fixed distance, and also if a fixed velocity were added to the velocities of all objects. He summarized this by stating that all motion is relative: *Motus inter corpora relativus tantum est*. From this symmetry it follows that not circular motion, but straight-line motion with constant speed is 'natural'.

The symmetries that govern the quantum world are mathematically similar to rotations. Quantum symmetries are like rotations in some abstract 'internal space'. A symmetry of that type is called *internal symmetry* to distinguish it from symmetries that are visible in space-time, such as spherical symmetry. Maybe this internal space is somehow related to the 'external space' of space-time, but at present this is entirely unclear.

By way of analogy, imagine that every coin in your wallet was rotated about its centre.[81] The weight of your wallet would not change, and the value of your money is still the same, even if you cannot look inside your wallet. Only the relative orientation of the coins would be different.

Mathematicians classify symmetries in *symmetry groups*. A symmetry group is determined by a set of rules that prescribe how objects can be 'changed so that nothing changes'. These rules are distant relatives of the 'Gang of Four', that is, the symbols + − × ÷ from ordinary arithmetic. For example, thinking back to Galileo's argument about 'natural motion': a circle does not change when rotated about its centre. In other

81 Many countries have coins that are not circularly symmetric, but have the form of a polygon. In that case, mathematicians speak of a *discrete symmetry*.

words, a circle is symmetric under the *rotation group* in two dimensions. A straight line does not change when it is displaced along itself; it is symmetric under the *translation group*.

Physicists have discovered three types of *internal symmetry* of particles. The mathematical family names of the symmetries that rule the particle world are U(1), SU(2) and SU(3). Why these, and not others, is still unknown.

That a symmetry can determine the interaction between objects was illustrated above in the case of two colliding balls. The consequences of an invisible internal structure of such balls may be demonstrated in an experiment that Huygens proposed in his *Traité de la Lumière*.[82] Suppose that we have a row of identical balls, suspended to form a line. The leftmost ball is pulled back and released. What will happen?

Symmetry provides the answer. Just walk around to the other side of the instrument. Because of the symmetry of space, what you see is the same as what you had before, except that now the rightmost ball is up in the air. Thus the prediction is that after the collision, the rightmost ball will swing upward to the same height at which the leftmost ball was released.

However, if the balls are not made of steel but of cork, or modeling clay, the outcome is drastically different. The symmetry is broken by the internal structure of the balls, which has observable consequences on a scale much larger than that of the material in the balls.

82 Written around 1660, published in 1690. Huygens's work on colliding objects was published only after his death: *De motu corporum ex percussione*, in *Opuscula Posthuma* (1703). The experiment is sold in scientific toy shops, often rather unfairly called 'Newton's cradle'.

→ Whether the image at the top shows the position before the collision, and the one at the bottom the position thereafter, depends on one's point of view in space. Symmetry says that these two arrangements are in fact mechanically identical.

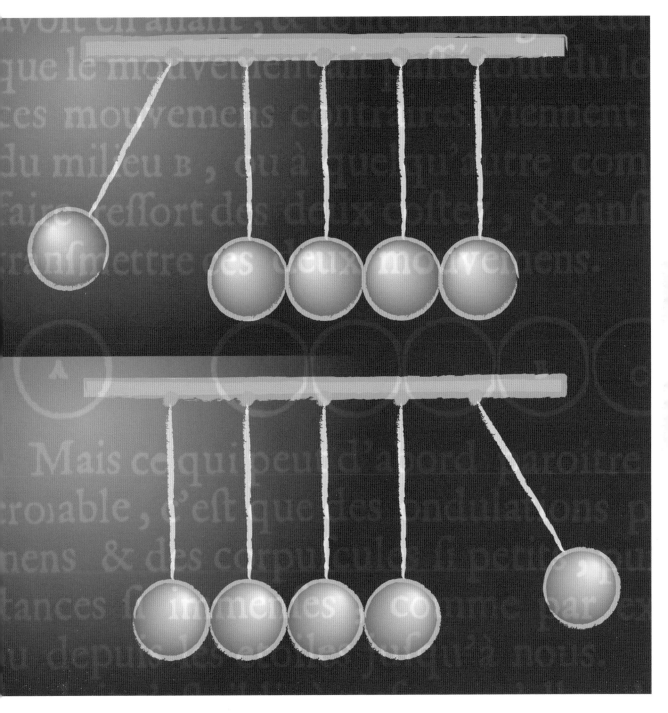

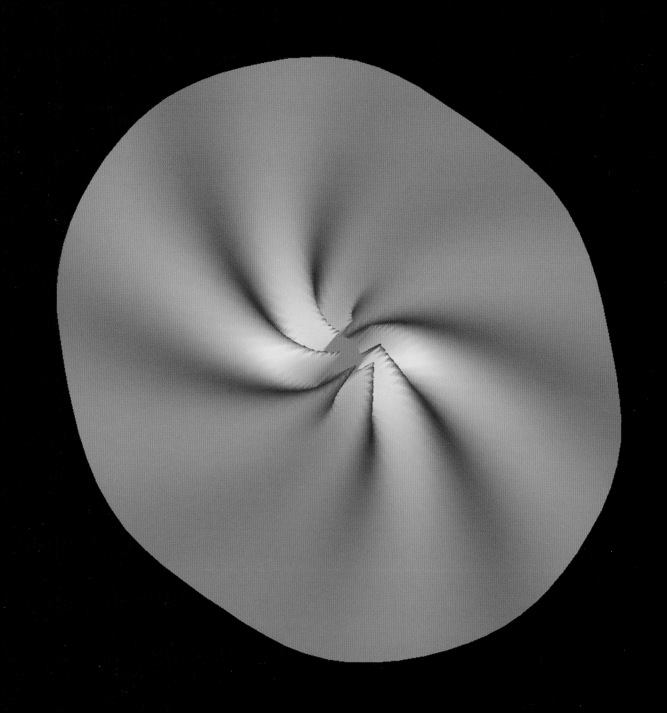

Speaking of 'broken symmetry': in the argument given above, the over-all symmetry of space is broken by the fact that the acceleration of gravity provides a specific observable direction.[83] 'Walking around the instrument' in this case amounts to a rotation about the vertical axis, the vertical being defined by the direction of the local gravity. Turning the row of balls around is fine; tilting it is not.

But wait. If a symmetry is a 'change that leaves something unchanged', it is unobservable. Just try it with a perfectly white billiards ball. How could a symmetry then produce the measurable effect of an interaction, as was claimed above? The answer to this question is as deep as it is surprising: because nothing can propagate faster than the speed of light, it is impossible to apply a symmetry to the whole Universe every-where at once. In other words: a *global symmetry is impossible*.

A simple tabletop experiment shows the consequences of this impos-sibility. Just as in the case of gravity, what goes on may be understood by means of an analogy in a two-dimensional world. The similarity be-tween the symmetries of the Universe and simple rotations enables us to understand how a symmetry can produce a force.[73]

Please perform the following experiment. It takes no more than two minutes, whereas it may well take two years to master the mathematics necessary to describe the quantum field theory that describes the way our world works.

← A local rotation makes a tablecloth wrinkle. The cloth is not symmetric under local rotations: there is a mismatch between the local twist in the middle and the undisturbed cloth further out. The mismatch shows up as wrinkles. These correspond to the appearance of a field (field lines) in the case of a local quantum symmetry. In a Feynman diagram, the wrinkle field shows up in the form of a boson at a vertex.

83 Yes, I know that it doesn't exist, but let's keep the name for household purposes. 'Temperature' doesn't exist either, but it's a bit silly to say, 'The mean energy per water molecule is too high', instead of 'The shower is too hot'.

Take a small tablecloth. The material must be a uniform colour, without any patterns. Make sure it is quite smooth (remember how to use an iron?) We are not looking so closely that the individual fibres are visible, and we will pretend that the material extends to infinity: this represents our world in two-dimensional analogy, just as did the paper with which we demonstrated the consequences of space curvature.

Now rotate the whole cloth through an arbitrary angle. The tablecloth has exactly the same appearance as before the rotation (remember, we imagine that the edges are all the way at infinity). Mathematicians say that the space represented by the fabric is *invariant under global rotations*. This part of the experiment seems thoroughly trivial, until you realize that it would be impossible, even in principle, to do this with the real Universe. Imagine that we want to perform a global symmetry transformation. Then we would have to let the symmetry act in all of space at exactly the same time. But it is impossible to arrange this in reality because no signal can propagate faster than the speed of light.

The speed of light is always the same. The symmetry that describes this invariance is called *Lorentz symmetry*, after its discoverer. Einstein proved that it then follows that the speed of light is also the maximum speed in our Universe. Thus, every signal takes time to cross space.

Einstein's relativity, based on Lorentz symmetry, tells us that we must abandon the idea that a symmetry can be global. Because nothing can travel faster than light, we cannot apply a symmetry in the whole Universe at once. The way out of this difficulty is to *accept only local symmetry rotations*, that is, a symmetry where the amount of rotation *differs* from event to event in space-time.

Return to the tablecloth before you. Put your finger on a point near the centre and give the cloth an arbitrary twist, keeping the edges of the cloth in place. In the vicinity of the twisted point, something has happened: *a spray of wrinkles radiates outward*. The local twist cannot be connected smoothly with the undisturbed cloth at large distances. The difference is patched up by the wrinkles.

Because of relativity, all symmetries must be local, and *any local symmetry creates a field*. The wrinkles are related to the *field lines*, as they were called by Faraday in pre-quantum days.

Again: a *local* symmetry is one that differs from point to point in space-time, or from point to point in some abstract general or internal space, maybe even in a 'super space-time' with more than the obvious (3+1) dimensions. A local symmetry twist creates a mismatch between the local area and the rest of the Universe. The mismatch is patched up by the appearance of a field.

That field shows up in our large-scale world as a force. In the quantum world the field is made of particles. A field particle is a boson, the kind of 'messenger particle' represented by the horizontal line in the H-shape of the Feynman diagram. In the analogy given above, the field boson is like a ball exchanged between two tennis players floating in free space, changing the state of motion of the players due to the recoil at each stroke of the racket.

Every symmetry corresponds to a specific type of field boson. Local U(1) symmetry creates a *photon* at a vertex, and thereby produces the forces of electromagnetism. Local SU(2) symmetry generates the W^+, W^- and Z bosons of the 'weak force', responsible for some types of radioactivity. And local SU(3) symmetry produces *gluons* that carry the whimsically named 'colour force', which couples quarks together to form protons, neutrons and atomic nuclei. From the Feynman diagrams and related expressions it may be computed how the corresponding forces depend on the distance in space.[84]

84 This is a very complicated issue. Roughly speaking, the electrostatic force behaves like gravity, with the force decreasing as the inverse square of the distance. The weak force behaves like the electrostatic force at small distances, but cuts off sharply when the particles are far apart. The colour force is small at close range, and increases indefinitely when the distances between the quarks become bigger.

The tablecloth analogy may help to understand an old problem: if an atom drops to a lower energy state and emits a photon, where was the photon before that? The answer is: the photon was in 'another world', an 'internal space' with a different orientation. The photon appeared when the U(1) twist gave access to the 'internal space' of the electron in the atom.

In the present state of theoretical physics, such symmetries are the all-encompassing organizing principle in the Universe. They dictate the way in which fermions and bosons are coupled at a vertex. Symmetries prescribe the interaction of matter (fermions) with forces (bosons). The remarkable thing is that *all known quantum interactions follow this pattern*. Symmetries are so powerful that, given a set of fermions, we can calculate exactly how many bosons can interact with them, and almost exactly what the properties of these bosons are.

Moreover, general relativity is, upon closer inspection, also a local-symmetry theory. Remember that a symmetry is a 'change that leaves something unchanged'. The dramatic result of the 1887 Michelson-Morley experiment was that the speed of light is always the same, independent of the state of motion of the emitter or the receiver. The Lorentz-symmetric theory of motion is Einstein's Special Theory of Relativity, in which space-time has the same structure everywhere at all times.

Extending this to *local* Lorentz symmetry, Einstein obtained the General Theory of Relativity, which is his special relativity modified to allow the structure of space-time to differ between events. In this way, the theory describes the structure and dynamics of space-time.

The descriptions of gravity and of quantum fields are all theories based on the application of local symmetry. Therefore, physicists have conceived the hope that a single explanation for all interactions will soon be discovered. In the case of quanta, many attempts at doing this have been made. These have names like 'grand unification', 'SU(5)', 'string theory' and so forth. Experiments with particle accelerators, and evidence from astrophysics, have ruled out all of the proposed 'super' theories so far. But many physicists think that in the early Universe all quantum interactions were due to a single, all-encompassing local symmetry.

Questions for the 21st Century

Where are we now, having covered four-and-a-quarter centuries of physics? At every point, what was the key observation or experiment? What was the problem associated with that? How did it get solved?

In 1585, Stevin performed his leaden ball experiment in Delft. The associated problems were the pronouncements inherited from Aristotle on falling objects. Stevin, a hero in the vanguard of 'experimental philosophy', was not impressed (or, in any case, not deterred) by this ancient stuff.

The solution of the problems surrounding falling objects was reached tactically by 'separation of difficulties'. On one front, Galileo performed his lengthy studies of 'natural' motion by means of clever observations and experiments on objects falling freely or rolling along inclined planes. That approach was concluded by Huygens's radical statement on the relativity of motion. From this he derived the rules for collisions and the first mathematical equations in theoretical physics: the centrifugal acceleration, the oscillation period of the pendulum, fall along curved paths such as the famous cycloid, and more.

The other front focused on the origin and description of the acceleration in free fall. Newton rehabilitated the shaky concept of 'force' by casting it in the strict mathematical form of 'universal gravity' and devised a brilliant formalism for calculating the results of accelerations (a technique we now call *differential and integral calculus*, invented independently by Leibniz).

The success of this approach was phenomenal, but it left all questions about the origin of forces and the structure of matter unanswered. Matter cannot be stable if it is subject to gravity only, so other ingredients were needed. Chemistry provided some clues through the discovery of chemical elements and the fact that these combine in whole-number ratio. H_2O is a chemical compound, where $H_{1.8}O$ would be just a mixture. Research into electromagnetism also offered some hope of understanding the inner forces of matter.

Maxwell's discovery that electromagnetic waves all propagate with the speed of light led to the 1887 Michelson-Morley experiment. This landmark observation showed that the speed of light does not depend on the motion of the emitter or the receiver of the light: the speed of light is invariant. This formed the basis of Einstein's Special Theory of Relativity in 1905, with, as a great bonus, the discovery that inertia is a consequence of the fact that the speed of light is a universal maximum speed.

Einstein followed this up in 1916 with an even greater milestone: the General Theory of Relativity, thereby doing away with gravity altogether. The theory provided an explanation of Stevin's demonstration that the acceleration of 'gravity' is independent of the falling object's mass. The success of Einstein's theories became literally enormous when the application of general relativity to the cosmos gave a coherent description of the overall structure and dynamics of the Universe.[85] Cosmology became a respectable branch of astronomy at last.

Almost at the same time, a large body of experimental results and associated questions reached such proportions that ignoring or finessing them was no longer an option. Chemistry, thermodynamics, crystallography, radiation from hot objects and radioactivity all indicated that the structure of matter was far more complicated than physicists had hoped. Rutherford and his co-workers found that atoms are made of a positively charged nucleus, attended by negatively charged electrons. Maxwell's theory of electromagnetism said that such an arrangement would vanish in a flash, yet lead was as stable in 1910 as it was in 1585.

The problem of the stability of matter was circumvented by Bohr, who

85 This is known as 'Big Bang theory', which is an awful misnomer, because it suggests that matter is moving through space after an explosion in one specific location. Actually, Einstein's theory describes *the motion of space itself*, and not *a motion somewhere in space*. Moreover, the name suggests an extremely violent process, while actually the initial Universe was in almost perfect thermal equilibrium. Besides, it is hardly 'just a theory', because the astronomical evidence for the expansion of space on a large scale is overwhelming.

guessed that atoms can only gain or lose energy in discrete portions, just like Planck had proposed for radiation in 1900. With the bravery and honesty of the great scientist, Bohr admitted to the 1912 Solvay Conference on theoretical physics that he had given a theory that reproduced the static energy states of the hydrogen atom, not its dynamical behaviour. This is comparable to the work of Archimedes and Stevin, who formulated the laws of mechanical equilibrium without having a theory of the motion of objects. Bohr had no dynamical theory at hand that explained his success. Such a theory, soon called *quantum mechanics*, emerged with the work of Schrödinger in 1926.

The rock bottom of contemporary physics can now be summarized as: *the Universe is built of quanta and space-time*. As simple as it sounds, this sentence is so explosive as to be practically self-destructive. Particles and space-time have to coexist somehow in order to make our Universe what it is. But how could that be? Particles are quanta, matter is grainy; but space-time is a smooth continuum. The Einstein equation describing the structure and dynamics of space-time is classical in the sense that a given past inexorably produces a specific future. Particles, on the other hand, obey quantum rules, in which both past and future must be specified before we may compute how probable the transition between the two is.

The most generally inclusive formulation of particles-and-space-time is still embodied in the Einstein equation.[53] Even though the quantum behaviour of matter is not included in the description, it has the enormous virtue of including space-time and matter on an equal footing. In all other theories, from classical mechanics right through quantum field theory, space-time is a passive stage on which the particle play is performed. We recall that the Einstein equation can be read as: *the structure and dynamics of space-time are determined by the arrangement of mass, energy and momentum.* This connection between matter and space-time is so evident as to be almost obvious. Experience shows that the curvature of space-time (also known as 'the strength of gravity') is always the strongest in regions where the largest amount of mass is concentrated.

This is remarkable, because the Einstein equation allows space-time to provide all the curvature entirely by itself without the presence of matter, as in the case of gravitational waves and black holes. Then what does matter have to do with it? How are Stevin's leaden balls connected to the gravity-smoothed sphere of Earth?

Einstein developed the right-hand side of his equation ('the arrangement of mass, energy and momentum') by analogy with an expression in Newtonian mechanics called Poisson's equation.[86] As brilliant as this was, it does not specify a physical connection between matter and space-time. Merely saying 'is determined by' is not explicit enough to count as the description of an actual mechanism.

The '=' sign in the Einstein equation apparently hides an as-yet-undiscovered coupling mechanism. It is rendered here as 'determined by'. What is that supposed to mean? Are we back in the days of Aristotle? What is the mechanism, the coupling, the interaction?

The physics waters are troubled yet further by a large number of contradictions, problems and dark hints, many of which are due to the application of quantum field theory and the Einstein equation to cosmology. Listing those and their connections would take a whole extra book. Besides, many of the astrophysical data and computations are shaky, never mind claims about 'precision cosmology'.

The most ominous of these problems is the discovery that the vast majority of the contents of our Universe is something else than matter of the type dropped by Stevin off that tower in 1585. The ordinary

86 Siméon Denis Poisson (1781-1840), French mathematician.

chemical elements taken together make up at most 5% of matter. All we know about the other 95% is that it produces space-time curvature ('gravity') and that it does not emit any light. Thus, it is usually called 'dark matter' and 'dark energy', but that is an overstatement because we don't even know that it is matter or energy in the usual sense.[87] The expression *dark stuff* would be better, for the time being.

Just as in a mystery novel, we will probably find that, once the case is closed, many or even most of the clues will turn out to have been so many red herrings. Right up to the final chapter the mystery remains, in our case summarized by: *How does the Sun curve space-time?* Likewise, Newton had to face the question: How does gravity produce a force? Where's the pulling string?

The predictions of general relativity have been verified with impressive precision. Matter responds to curved space-time by following curved orbits. Space-time responds to the presence of matter by assuming a curvature. There is a reciprocity there, a coupling, but how does it work?

What is the interaction between quanta and space-time?

In quantum field theory, interaction is mediated by a particle, a 'field boson'. But then space-time itself should have some kind of micro-

87 Technical detail: astronomers find that the best correspondence between theory and observations is obtained if about two-thirds of the dark stuff is 'dark energy' and one-third is 'dark matter'. But it follows from the Einstein equation that, during the evolution of the Universe, the balance shifts from almost all 'dark matter' to almost all 'dark energy'. I find it too much of a coincidence that, right now when we naked apes come along with our telescopes, the division should be about fifty-fifty.

→ At a vertex (white) a fermion (brown) couples to a boson (blue). This three-way connection is the basic building block for all Feynman diagrams belonging to a given symmetry. By adding two vertices, the boson temporarily transforms into a fermion-antifermion pair. In this way, an infinity of increasingly complex diagrams can be built. The net effect of the interaction is due to the sum over all possible diagrams.

structure. There ought to be some kind of field particle that connects material particles in such a way that, seen on our very large length scales, the resulting force behaves like Einstein's curved space-time.

The alternative is that the particle structure of matter smoothes out when we look at still-smaller length scales: electrons and their friends lose their identity and all of them become smooth under yet more powerful magnification. That possibility is so far out, and so without precedent, that it does not seem to provide a productive approach at this moment.

The first option calls for some kind of super Feynman diagram in which the space-time field is built of intermediary particles, pre-emptively called 'gravitons'. But it has been found that naively transplanting the quantum formalism to the space-time domain produces immense problems that seem insoluble at present.

The snag is the following. The H-shape is only the simplest of all possible Feynman diagrams. There are a multiple infinity of 'higher order' diagrams, all of which have more than the minimum two vertices. For example, the intermediary boson — the horizontal line of the H — may split locally into a fermion-antifermion pair. In our large-scale world, this pair is not seen directly. Only its indirect consequences are observed, in the form of certain properties of the electromagnetic force.

Stranger still, this additional pair of particles may appear spontaneously in space near the field boson. In a local-symmetry quantum field theory, the number of particles participating in an interaction is

→ In this example, a fermion-antifermion pair appears spontaneously. Feynman diagrams of this type are perfectly legitimate. Because of quantum behaviour, the number of particles in a volume of space is not constant.

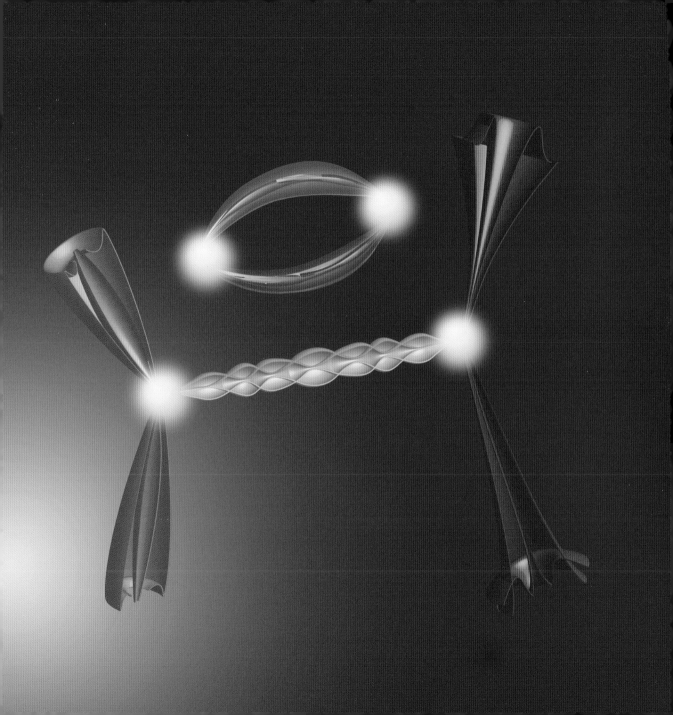

not fixed. That is dramatically different from our large world. Even if a snooker player pockets some balls, nobody thinks that they have truly disappeared, but the rules of quantum snooker do not call for a fixed number of particles.

Thus, we find that in relativistic quantum field theory, the vacuum is not really devoid of particles. The effect is called *vacuum polarization*. In fact, we cannot ever say whether or not a given cubic metre of space is empty or not. And therein lies the problem: in the Einstein equation, the structure and dynamics of space-time are determined by the distribution of mass, energy and momentum. Therefore, the spontaneously appearing and disappearing vacuum particle pairs should contribute to the strength of the resulting 'gravity'.

This phenomenon is called the *Casimir-Polder force* after its discoverers.[88] It is not too difficult to estimate how strong this effect is. For comparison, we may express this force by computing what the density of matter would have to be if it were to have the same effect as the vacuum polarization.

If all of 'empty' space (the technical term, oddly enough, is still *vacuum*) were indeed pervaded by virtual particles, then the vacuum would behave as if it had a stupendous mass-energy density. For the U(1) and SU(2) symmetries, that would be 10^{34} times the cosmic density in stars and galaxies. For the SU(3), the factor would be as large as 10^{118}, that is, 1 followed by 118 zeros. From astronomical data we know that this is out of the question: the Universe would vanish almost instantly in a flash.

That brings us to the Big Question of our times. Either the Feynman diagram of the vacuum particle-antiparticle pairs must be read in a different way, or it must be extended to explicitly include the coupling between those particles and space-time. But how?

88 Hendrik Brugt Gerhard Casimir (1909-2000) and Dirk Polder (1919-2001), Dutch physicists: 'The Influence of Retardation on the London-Van der Waals Forces', *Physical Review* 73, pp. 360-372, 1948.

Small Moves, Ellie

During one of my visits to CERN, the European research institute for particle physics in Geneva, I met Elmajid Nath-Kaci-Uvutmar, a Touareg from North Africa. As a boy, he fell seriously ill and was nursed back to health by the 'White Fathers' in a desert monastery. They renamed him Majid Boutemeur while he stayed there for his education until he was about 14 years old. He found work on a small boat, cleaning fish, until one day in Marseille he decided to stay in France to be trained in physics. I met him at CERN where he was working on the development of software for particle accelerators. I asked him: 'You were fourteen when you decided to become a physicist. Suppose that you meet a thirteen- or fourteen-year-old boy or girl, asking you: please give me one good reason to go and study physics. What would you reply?' He looked into my eyes and said:

> *Physics is the most wonderful thing. You should be so lucky to go home in the evening and still have a problem to solve.*

Brilliant, the true grit of the scientist knowing that the primary product of research is failure. *'The conditions of nature give us no alternative'*, Huygens warned us. Even a 'failed' experiment or theory is useful when mapping new land. How awful would it be if I were to come home one evening and say to my wife and daughter: 'Darlings, physics is finished.' Fortunately, such a disaster will never happen. *We should be so lucky*.

That which we call 'gravity' is one of the consequences of the structure of space-time. Einstein asserted that the mass-energy-momentum of matter contributes to that structure. But how? What is the interaction? What mechanism is hidden in the = of the Einstein equation? Or, more plainly: how does the Sun tell space-time around it that it's supposed to be curved, and by how much?

The present generation of physicists hasn't found an answer yet, or even a formulation that is clearly a good starting point. Possibly younger folk must give it a go. I hope that they will not have the wrong image of advancement in the sciences. Contrary to street wisdom, great advances in physics are not revolutionary but stepwise. Einstein's life and achievements have been romanticized beyond recognition, but he wasn't a rebel or iconoclast. He had the unique combination of perceptiveness, knowledge and intelligence not to think that something is 'evidently' impossible; witness his acceptance of the invariance of the speed of light, and Planck's quanta.

Science does not proceed after the manner of human history or modern architecture: slash and burn, blast and demolish, and on the rubble build something that doesn't resemble the past in any way. A good scientist is more like a good dentist: remove what cannot be maintained, leave the rest as it is, install replacements, and improve the functioning of the whole.

That is what makes science especially difficult. A proper new theory includes the old as a special case or an approximation. Classical mechanics is included in special relativity, if all velocities are small compared with the speed of light. Nuclear and atomic physics are included in quantum field theory, if the objects are much bigger and slower than photons or quarks. Chemistry is atomic physics, but at energies that

→ Nobody knows what images will illustrate the theories of the future. Maybe someday a picture like this one will illustrate a textbook on space-time-quantum theory.

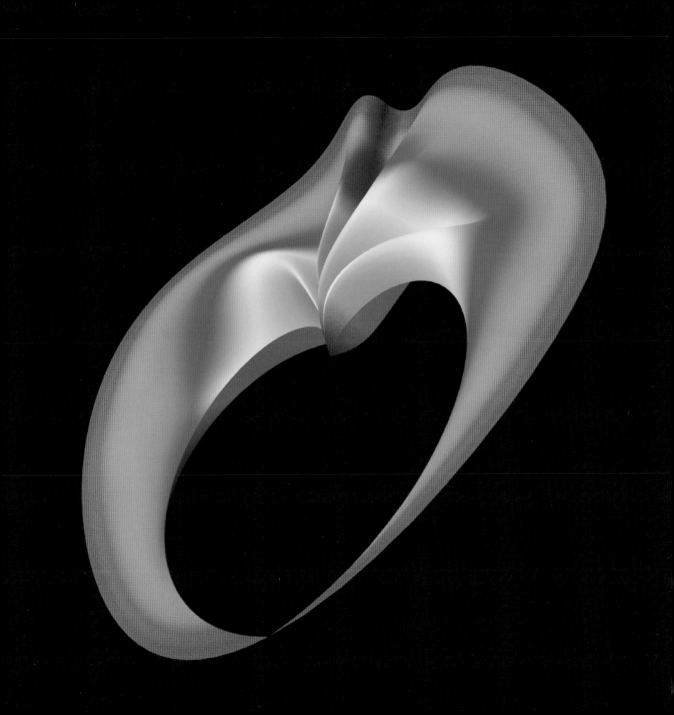

are much smaller than those of the atomic nuclei. The necessity to fit all such formulations together makes building a good theory very hard indeed.

Nor does science advance with enormous leaps and bounds. It is not wise to take a hundred-kilometre jump seaward when standing on the shore of an ocean of unknowns. Sagan describes in his book, *Contact,* how the heroine, as a child, tells her father what enormous strides she intends to make in science.[89] Daddy warns her: 'Small moves, Ellie. Small moves.' She heeds the lesson, and becomes the first person on Earth to make contact with an extraterrestrial civilization.

How does science proceed, then? There is no method, no *méthode*.[26] What can we do to help our next generation? My recommendation is: be perceptive, aware, attentive. About what? That is anybody's guess. Consider: there are three families of elementary particles, and three space dimensions. Coincidence, or road sign? Consider: quantum interactions are determined by symmetries in an abstract 'internal space'. What space? Is it related to the 'external space' of space-time? Consider: dark stuff dominates the Universe, and roughly half of it is 'matter' and the other half 'energy'. Coincidence, or road sign? Consider... well, what?

The present generation of theorists seems to have skipped the 'perceptiveness' stage of scientific inquiry, and has gone straight to 'hypothesis' with a display of mathematical effort that grossly intimidates most people, including many physicists.[90] People should read history

89 Carl Edward Sagan (1934-1996), American astronomer: *Contact*, Simon & Schuster, New York 1985. I also warmly recommend the movie, with Jodie Foster starring as Ellie.

90 Some of my colleagues in pure mathematics do not find these theories quite so impressive. I suppose that the future will have to pass judgment, as usual.

→ Classical orbits are curved due to an acceleration; space-time structure provides curvature all by itself; local symmetries govern the interaction between quanta; but what insight will merge all these into a single theory?

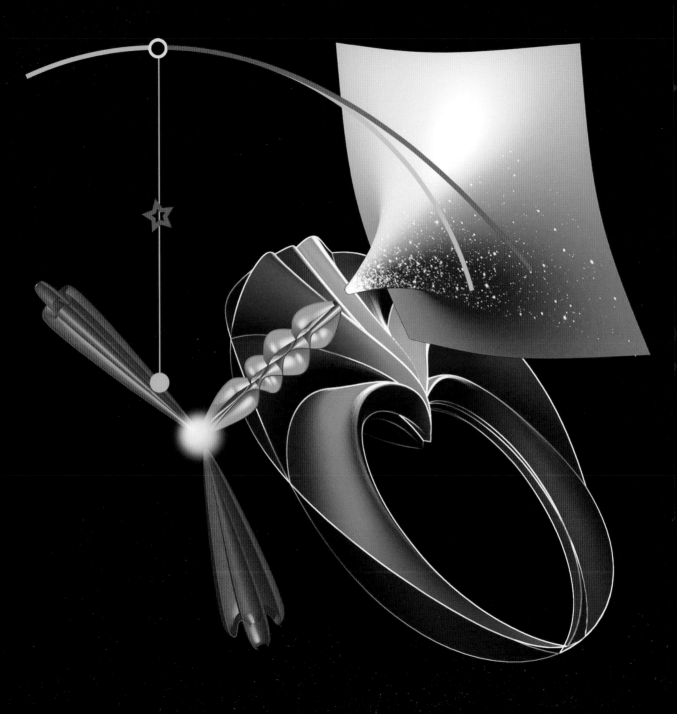

before they go out on a limb. Greek philosophy more or less floated on hypotheses, posited as self-evident and immutable truths, of course. What to do? Going around with a lantern in broad daylight, like Diogenes,[91] will probably not have the desired effect.

Being an old scientist is a mixed blessing when confronted with the enormous task of solving the interaction between particles and space-time. It is good to have a large store of knowledge, to serve as a survival kit in inhospitable territory. But it is bad to be straitjacketed by it. Small moves is right, but do not think too soon that something cannot be. As the White Queen said to Alice in Wonderland: *'Sometimes I've believed as many as six impossible things before breakfast.'* Creating knowledge should take precedence over having it.

I do not want to leave the impression that this huge task is something to be sad about. Quite the contrary: in physics, big problems are the heralds of big findings. Discoveries spring eternal, and maybe the answer (or the right question) is nearby — even obvious, later generations might say with the perfect vision of their hindsight. Schopenhauer wrote to Darwin:[92]

> *Our task is not: to see what nobody has seen yet, but: to think what nobody has ever thought, about things that everyone sees.*

One thing we know with absolute certainty, after so many centuries of physics: we're into *experimental* philosophy, so we have work to do if we want to solve that big question. We even know more or less in what direction we should look for the outlines of answers. Plural, of course, because four centuries of history have shown that nothing in physics is simple.

91 Diogenes of Sinope (412-323), Greek philosopher.
92 Arthur Schopenhauer (1788-1860), German philosopher; Charles Robert Darwin (1809-1882), English biologist.

How does the Sun curve space-time around itself? Sounds good, but it is both too general and too specific to be a proper physics question. In 1987, the rock band U2 had a hit song on their album *The Joshua Tree* containing the refrain:

And I still haven't found what I'm looking for.

When that happens, it is wise to take a break, step back, and stop asking pointed questions. In the time between the year 1585 and now, we have learned that that usually helps.

Thanks

Contrary to what the cynics may say, many non-scientists are captivated by the same Big Questions that keep physicists and astronomers so busy. This book is an attempt to share my fascination with them. I am grateful to the people at Amsterdam University Press for encouraging me to do so. Even though this book is primarily meant for the general public, I wish to hold it to the professional standards of a scholarly work. I am most grateful to six scientists for the time and effort they spent in assessing an earlier version of my book. It is customary that such referees remain anonymous, so that they are entirely free to tell the author their opinions. Therefore, I cannot thank them in person for their criticism and insight, but my gratitude remains undiminished nonetheless.

Fortunately, the seventh referee, Charlotte Lemmens, did not remain anonymous at all, which gives me the joyful opportunity to thank her once more for her detailed and intelligent assessment. I hope that she may consider this result as yet another bright feather in her guardian angel wings.